# Pharmaceutical Crystals

# Pharmaceutical Crystals

Special Issue Editors

**Etsuo Yonemochi**
**Hidehiro Uekusa**

MDPI • Basel • Beijing • Wuhan • Barcelona • Belgrade • Manchester • Tokyo • Cluj • Tianjin

*Special Issue Editors*

Etsuo Yonemochi
Hoshi University
Japan

Hidehiro Uekusa
Tokyo Institute of Technology
Japan

*Editorial Office*
MDPI
St. Alban-Anlage 66
4052 Basel, Switzerland

This is a reprint of articles from the Special Issue published online in the open access journal *Crystals* (ISSN 2073-4352) (available at: https://www.mdpi.com/journal/crystals/special_issues/pharmaceutical_crystals).

For citation purposes, cite each article independently as indicated on the article page online and as indicated below:

LastName, A.A.; LastName, B.B.; LastName, C.C. Article Title. *Journal Name* **Year**, *Article Number*, Page Range.

**ISBN 978-3-03928-712-3 (Pbk)**
**ISBN 978-3-03928-713-0 (PDF)**

Cover image courtesy of Hidehiro Uekusa.

© 2020 by the authors. Articles in this book are Open Access and distributed under the Creative Commons Attribution (CC BY) license, which allows users to download, copy and build upon published articles, as long as the author and publisher are properly credited, which ensures maximum dissemination and a wider impact of our publications.

The book as a whole is distributed by MDPI under the terms and conditions of the Creative Commons license CC BY-NC-ND.

# Contents

**About the Special Issue Editors** . . . . . . . . . . . . . . . . . . . . . . . . . . . . . . . . vii

**Etsuo Yonemochi and Hidehiro Uekusa**
Preface of the Special Issue "Pharmaceutical Crystals"
Reprinted from: *Crystals* **2020**, *10*, 89, doi:10.3390/cryst10020089 . . . . . . . . . . . . . . . . . . . . 1

**Yan Ren, Jie Shen, Kaxi Yu, Chi Uyen Phan, Guanxi Chen, Jiyong Liu, Xiurong Hu and Jianyue Feng**
Impact of Crystal Habit on Solubility of Ticagrelor
Reprinted from: *Crystals* **2019**, *9*, 556, doi:10.3390/cryst9110556 . . . . . . . . . . . . . . . . . . . . 5

**Yan Zhang, Zhao Yang, Shuaihua Zhang and Xingtong Zhou**
Synthesis, Crystal Structure, and Solubility Analysis of a Famotidine Cocrystal
Reprinted from: *Crystals* **2019**, *9*, 360, doi:10.3390/cryst9070360 . . . . . . . . . . . . . . . . . . . . 21

**Reiko Yutani, Ryotaro Haku, Reiko Teraoka, Chisato Tode, Tatsuo Koide, Shuji Kitagawa, Toshiyasu Sakane and Toshiro Fukami**
Comparative Evaluation of the Photostability of Carbamazepine Polymorphs and Cocrystals
Reprinted from: *Crystals* **2019**, *9*, 553, doi:10.3390/cryst9110553 . . . . . . . . . . . . . . . . . . . . 31

**Eram Khan, Anuradha Shukla, Karnica Srivastava, Debraj Gangopadhyay, Khaled H. Assi, Poonam Tandon and Venu R. Vangala**
Structural and Reactivity Analyses of Nitrofurantoin–4-dimethylaminopyridine Salt Using Spectroscopic and Density Functional Theory Calculations
Reprinted from: *Crystals* **2019**, *9*, 413, doi:10.3390/cryst9080413 . . . . . . . . . . . . . . . . . . . . 43

**Ryo Mizoguchi and Hidehiro Uekusa**
Elucidation of the Crystal Structures and Dehydration Behaviors of Ondansetron Salts
Reprinted from: *Crystals* **2019**, *9*, 180, doi:10.3390/cryst9030180 . . . . . . . . . . . . . . . . . . . . 57

**Dan Du, Guo-Bin Ren, Ming-Hui Qi, Zhong Li and Xiao-Yong Xu**
Solvent-Mediated Polymorphic Transformation of Famoxadone from Form II to Form I in Several Mixed Solvent Systems
Reprinted from: *Crystals* **2019**, *9*, 161, doi:10.3390/cryst9030161 . . . . . . . . . . . . . . . . . . . . 73

**Yaohui Huang, Ling Zhou, Wenchao Yang, Yang Li, Yongfan Yang, Zaixiang Zhang, Chang Wang, Xia Zhang and Qiuxiang Yin**
Preparation of Theophylline-Benzoic Acid Cocrystal and On-Line Monitoring of Cocrystallization Process in Solution by Raman Spectroscopy
Reprinted from: *Crystals* **2019**, *9*, 329, doi:10.3390/cryst9070329 . . . . . . . . . . . . . . . . . . . . 87

**Hyunseon An, Insil Choi and Il Won Kim**
Melting Diagrams of Adefovir Dipivoxil and Dicarboxylic Acids: An Approach to Assess Cocrystal Compositions
Reprinted from: *Crystals* **2019**, *9*, 70, doi:10.3390/cryst9020070 . . . . . . . . . . . . . . . . . . . . 101

**Aleksandr V. Ivashchenko, Oleg D. Mitkin, Dmitry V. Kravchenko, Irina V. Kuznetsova, Sergiy M. Kovalenko, Natalya D. Bunyatyan and Thierry Langer**
Synthesis, X-Ray Crystal Structure, Hirshfeld Surface Analysis, and Molecular Docking Study of Novel Hepatitis B (HBV) Inhibitor: 8-Fluoro-5-(4-fluorobenzyl)-3-(2-methoxybenzyl)-3,5-dihydro-4H-pyrimido[5,4-b]indol-4-one
Reprinted from: *Crystals* **2019**, *9*, 379, doi:10.3390/cryst9080379 . . . . . . . . . . . . . . . . . . . . 109

Reem I. Al-Wabli, Alwah R. Al-Ghamdi, Suchindra Amma Vijayakumar Aswathy, Hazem A. Ghabbour, Mohamed H. Al-Agamy, Issac Hubert Joe and Mohamed I. Attia
**(2*E*)-2-[1-(1,3-Benzodioxol-5-yl)-3-(1*H*-imidazol-1-yl) propylidene]-*N*-(2-chlorophenyl)hydrazine carboxamide: Synthesis, X-ray Structure, Hirshfeld Surface Analysis, DFT Calculations, Molecular Docking and Antifungal Profile**
Reprinted from: *Crystals* **2019**, *9*, 82, doi:10.3390/cryst9020082 . . . . . . . . . . . . . . . . . . . **123**

# About the Special Issue Editors

**Etsuo Yonemochi**, Pharmacist and Professor of Department of Physical Chemistry at Hoshi University, was born in 1961. He graduated from the Faculty of Pharmaceutical Sciences, Chiba University (1985), and received Ph.D. in 1991. In 1987, he joined the Faculty of Pharmaceutical Sciences, Chiba University as a Research Associate, then he spent two years at the School of Pharmacy, University of London as a research fellow. He moved to Toho University as an Associate Professor in 1996 and moved to Hoshi University in 2013. His main fields of interest are characterization of pharmaceutical products and application of in silico simulation and various analytical method to pharmaceutical formulation. He is a Chairperson of the committee members of Analytical Methods in Japanese Pharmacopeia, a Vice-President of the Japan Society of Pharmaeutical Machinary and Engineering, and a Councilor of Academy of Pharmaceutical Science and Technology Japan. His hobby is racing his horses.

**Hidehiro Uekusa**, Associate Professor of Department of Chemistry at Tokyo Institute of Technology, was born in 1964 in Tokyo. He received B.S. in 1987, M.S. in1989, and Ph.D. in 1992 from Keio University. In 1992, he joined the Department of Chemistry at Tokyo Institute of Technology as a Research Associate (1992–1999), then was appointed to Associate Professor in 1999. His primary field is chemical crystallography. His current interests include pharmaceutical crystals and their phase transitions, analysis of crystalline state reactions, and crystal structure analysis from powder diffraction data. He stayed at the Department of Chemical Science at Birmingham University, U.K., in 2002 for collaborative work in these areas. He is a member of the Crystallographic Society of Japan and the International Union of Crystallography. He enjoys teatime every afternoon with his students.

*Editorial*

# Preface of the Special Issue "Pharmaceutical Crystals"

Etsuo Yonemochi [1,*] and Hidehiro Uekusa [2,*]

1. Department of Physical Chemistry, School of Pharmacy and Pharmaceutical Sciences, Hoshi University, 2-4-41 Ebara, Shinagawa-ku, Tokyo 142-8501, Japan
2. Department of Chemistry, School of Sciences, Tokyo Institute of Technology, 2-12-1 Ookayama, Meguro-ku, Tokyo 152-8551, Japan
* Correspondence: e-yonemochi@hoshi.ac.jp (E.Y.); uekusa@chem.titech.ac.jp (H.U.); Tel.: +81-3-5498-5048 (E.Y.); +81-3-5734-3529 (H.U.)

Received: 1 February 2020; Accepted: 2 February 2020; Published: 5 February 2020

---

Dear Colleagues,

We are delighted to deliver the "Pharmaceutical Crystals" Special Issue of *Crystals*.

The crystalline state is the most used and essential form of solid active pharmaceutical ingredients (APIs). The characterization of pharmaceutical crystals encompasses numerous scientific disciplines, and its center is crystal structure analysis, which reveals the molecular structure of important pharmaceutical compounds, This analysis also affords key structural information that relates to the broadly variable physicochemical properties of the APIs, such as solubility, stability, tablet ability, color, and hygroscopicity.

The Special Issue on "Pharmaceutical Crystals" aimed to publish novel molecular and crystal structures of pharmaceutical compounds, especially new crystal structures of APIs, including polymorphs and solvate crystals, as well as multi-component crystals of APIs such as co-crystals and salts. Although these pharmaceutical crystals have the same API, they may lead to different physicochemical properties depending on their unique structures.

Thus, this Special Issue demonstrates the importance of crystal structure information in many sectors of pharmaceutical science. Ten groups, both from industry and academia, contributed their latest studies that include morphology, spectroscopic, theoretical calculation, and thermal analysis with the crystallographic study. This wide variety of studies is the key to this Special Issue presenting current trends in the structure–property study of pharmaceutical crystals.

In this Special Issue, physicochemical properties and crystal structure are the focus, and a variety of properties were correlated to crystal structure. Solubility of a pharmaceutical crystal is one of the most exciting topics, and two groups contributed to this aspect. Ren et al. [1] studied the relationship between the solubility and crystal faces and crystal habits, providing a new idea on the mechanism of crystal habit modification and its impact on solubility. Co-crystal formation is known as one of the effective methods to improve solubility. Zhang et al. [2] found a novel co-crystal of the potent H2 receptor antagonist famotidine (FMT) with masonic acid, which was stable, and showed higher solubility than the intact crystalline phase. Another essential property of photostability was evaluated in carbamazepine polymorphs (forms I to III) and three co-crystals [3]. Yutani et al. fully utilized FT-IR, low-frequency Raman spectroscopy, and solid-state NMR to find that lower molecular mobility is the key to higher photostability. The chemical reactivity of pharmaceutical salt, nitrofurantoin–4-dimethylaminopyridine (NF-DMAP), was examined by Khan et al. [4] using DFT methods and spectroscopy to conclude that the API was chemically less reactive compared to the salt.

Dynamic phenomena such as dehydration phase transformation and solvent-mediated phase transformation are also an important aspect of pharmaceutical crystals because they relate to the stability of crystals. Dehydration behavior of ondansetron hydrochloride and hydrobromide was reported by Mizoguchi et al. [5] to elucidate the mechanism. They utilized a recently developed

"Structure Determination from Powder Diffraction Data (SDPD)" technique to analyze the dehydrated crystal structures. Solvent-mediated polymorphic transformation of famoxadone from form II to form I was disclosed by Du et al. [6]. The transformation process was monitored by process analytical technologies and was found to be controlled by form I growth. It is interesting that hydrogen-bonding ability and dipolar polarizability affected the transformation.

How do crystals grow? Huang et al. successfully monitored the co-crystallization process in solution by Raman spectroscopy [7]. The authors found that suspension density and temperature both have an impact on the co-crystal formation. Besides the crystal growth, identification of the co-crystal composition is the critical step of any further analysis. An et al. successfully utilized the melting diagrams for adefovir dipivoxil and dicarboxylic acids [8]. This method is powerful in assessing the co-crystal composition in solid-state crystallization.

Molecular docking is an emerging topic for pharmaceutical crystal study. Ivashchenko et al. reported the crystal structure of a new biologically active molecule, which was also investigated as a new inhibitor of hepatitis B in a molecular docking study [9]. This substance has in vitro nanomolar inhibitory activity against the hepatitis B virus (HBV). Another docking study was reported by Al-Wabli et al. [10], where a newly synthesized compound was crystallographically characterized. Furthermore, the structure was analyzed using molecular docking studies and Hirshfeld surface analysis. The in vitro antifungal potential of the compound was examined against four different fungal strains.

In conclusion, this Special Issue presents a wide range of recent studies about pharmaceutical crystals and provides valuable information for future studies in the related fields. The guest editors hope the readers enjoy this beneficial Special Issue of "Pharmaceutical Crystals".

Prof. Etsuo Yonemochi
Prof. Hidehiro Uekusa
*Guest Editors*

## References

1. Ren, Y.; Shen, J.; Yu, K.; Phan, C.U.; Chen, G.; Liu, J.; Hu, X.; Feng, J. Impact of Crystal Habit on Solubility of Ticagrelor. *Crystals* **2019**, *9*, 556. [CrossRef]
2. Zhang, Y.; Yang, Z.; Zhang, S.; Zhou, X. Synthesis, Crystal Structure, and Solubility Analysis of a Famotidine Cocrystal. *Crystals* **2019**, *9*, 360. [CrossRef]
3. Yutani, R.; Haku, R.; Teraoka, R.; Tode, C.; Koide, T.; Kitagawa, S.; Sakane, T.; Fukami, T. Comparative Evaluation of the Photostability of Carbamazepine Polymorphs and Co-crystals. *Crystals* **2019**, *9*, 553. [CrossRef]
4. Khan, E.; Shukla, A.; Srivastava, K.; Gangopadhyay, D.; Assi, K.H.; Tandon, P.; Vangala, V.R. Structural and Reactivity Analyses of Nitrofurantoin–4-dimethylaminopyridine Salt Using Spectroscopic and Density Functional Theory Calculations. *Crystals* **2019**, *9*, 413. [CrossRef]
5. Mizoguchi, R.; Uekusa, H. Elucidation of the Crystal Structures and Dehydration Behaviors of Ondansetron Salts. *Crystals* **2019**, *9*, 180. [CrossRef]
6. Du, D.; Ren, G.-B.; Qi, M.-H.; Li, Z.; Xu, X.-Y. Solvent-Mediated Polymorphic Transformation of Famoxadone from Form II to Form I in Several Mixed Solvent Systems. *Crystals* **2019**, *9*, 161. [CrossRef]
7. Huang, Y.; Zhou, L.; Yang, W.; Li, Y.; Yang, Y.; Zhang, Z.; Wang, C.; Zhang, X.; Yin, Q. Preparation of Theophylline-Benzoic Acid Cocrystal and On-Line Monitoring of Co-crystallization Process in Solution by Raman Spectroscopy. *Crystals* **2019**, *9*, 329. [CrossRef]
8. An, H.; Choi, I.; Kim, I.W. Melting Diagrams of Adefovir Dipivoxil and Dicarboxylic Acids: An Approach to Assess Co-crystal Compositions. *Crystals* **2019**, *9*, 70. [CrossRef]
9. Ivashchenko, A.V.; Mitkin, O.D.; Kravchenko, D.V.; Kuznetsova, I.V.; Kovalenko, S.M.; Bunyatyan, N.D.; Langer, T. Synthesis, X-Ray Crystal Structure, Hirshfeld Surface Analysis, and Molecular Docking Study of Novel Hepatitis B (HBV) Inhibitor: 8-Fluoro-5-(4-fluorobenzyl)-3-(2-methoxybenzyl)-3,5-dihydro-4H-pyrimido[5,4-b]indol-4-one. *Crystals* **2019**, *9*, 379. [CrossRef]

10. Al-Wabli, R.I.; Al-Ghamdi, A.R.; Aswathy, S.A.V.; Ghabbour, H.A.; Al-Agamy, M.H.; Hubert Joe, I.; Attia, M.I. (2E)-2-[1-(1,3-Benzodioxol-5-yl)-3-(1H-imidazol-1-yl)propylidene]-N-(2-chlorophenyl)hydrazine carboxamide: Synthesis, X-ray Structure, Hirshfeld Surface Analysis, DFT Calculations, Molecular Docking and Antifungal Profile. *Crystals* **2019**, *9*, 82. [CrossRef]

© 2020 by the authors. Licensee MDPI, Basel, Switzerland. This article is an open access article distributed under the terms and conditions of the Creative Commons Attribution (CC BY) license (http://creativecommons.org/licenses/by/4.0/).

Article
# Impact of Crystal Habit on Solubility of Ticagrelor

Yan Ren, Jie Shen, Kaxi Yu, Chi Uyen Phan, Guanxi Chen, Jiyong Liu, Xiurong Hu * and Jianyue Feng *

Department of Chemistry, Zhejiang University, Hangzhou 310028, China; renyan0916@zju.edu.cn (Y.R.); shenjie1003@zju.edu.cn (J.S.); 21837073@zju.edu.cn (K.Y.); Pha3409@zju.edu.cn (C.U.P.); guanxi@zju.edu.cn (G.C.); liujy@zju.edu.cn (J.L.)
* Correspondence: huxiurong@zju.edu.cn (X.H.); jyfeng@zju.edu.cn (J.F.)

Received: 17 September 2019; Accepted: 23 October 2019; Published: 24 October 2019

**Abstract:** Drugs with poor biopharmaceutical performance are the main obstacle to the development and design of medicinal preparations. The anisotropic surface chemistry of different surfaces on the crystal influences its physical and chemical properties, such as solubility, tableting, etc. In this study, the antisolvent crystallization and rapid-cooling crystallization were carried out to tune the crystal habits of ticagrelor (TICA) form II. Different crystal habits of ticagrelor (TICA) form II (TICA-A, TICA-B, TICA-C, TICA-D, and TICA-E) were prepared and evaluated for solubility. The single-crystal diffraction (SXRD) indicated that TICA form II belongs to the triclinic P1 space group with four TICA molecules in the asymmetric unit. The TICA molecules are generated through intermolecular hydrogen bonds along the (010) direction, forming an infinite molecular chain, which are further stacked by hydrogen bonds between hydroxyethoxy side chains, forming molecular circles composed of six TICA molecules along bc directions. Thus, in the case of TICA form II, hydrogen bonds drive growth along one axis (b-axis), which results in the formation of mostly needle-shape crystals. Morphology and face indexation reveals that (001), (010) and (01-1) are the main crystal planes. Powder diffractions showed that five habits have the same crystal structure and different relative intensity of diffraction peak. The solubility of the obtained crystals showed the crystal habits affect their solubility. This work is helpful for studying the mechanism of crystal habit modification and its effect on solubility.

**Keywords:** ticagrelor; crystal structure; crystal habit; solubility; dissolution

---

## 1. Introduction

Ticagrelor (TICA) is an oral antiplatelet drug that can be used in combination with a small amount of aspirin to reduce the danger of stroke and myocardial infarction in patients with acute coronary syndrome [1–3]. Similar to thiophene pyridine, ticagrelor inhibits the pro-thrombotic effect of ADP by blocking the platelet P2Y12 receptor. Unlike the thieno pyridines, ticagrelor reversibly binds to the P2Y12 receptor, showing rapid onset and offset of effect and does not require metabolic activation [4]. TICA has been used in clinical trials to reduce the incidence of recurrent myocardial infarction and stent thrombosis and was approved for use in the USA in 2011 [5–10]. According to the patent, TICA presents four polymorphisms (I, II, III and IV) and several pseudopolymorphs, such as monohydrate and DMSO solvate. However, only two crystal structures of them (form I and DMSO solvate) have been reported [11–13]. Different crystal forms have different stability, solubility, fluidity, etc., among which the TICA form II has the best stability, so it is widely used in clinical and has great commercial value. Unfortunately, ticagrelor belongs to biopharmaceutics classification system(BCS) class IV drug, with limited bioavailability (30–42%) [14].

Improving the dissolution rate is the key to obtaining a therapeutic effect and the rate-limiting step for bioavailability. The solubility and bioavailability are generally improved by crystal characteristics such as crystal habit, polymorphism and reduction of the particle size [15–20]. There have been many studies demonstrating the effect of polymorphism on oral bioavailability and/or dissolution rate [21]. However, the dissolution rate not only differs for different polymorphisms, but also, for different crystal habits [22], which has received scant attention. Meanwhile, crystal habits also influence stability, flowability, suspension, packing, density, compaction, etc. [23–27]. Thus, optimizing crystal properties by modification of the crystal habit of a drug seems to offer an alternative approach to changing the bioavailability of drugs. The relative growth rate of each surface determines the overall shape of the crystal. The growth rate of the crystal surface will be controlled by a combination of structure-related factors, such as dislocations and intermolecular bonds, and by exterior factors such as solvents, rate of agitation, additives, temperature, etc. [28–35].

This study aims to systematically investigate how crystal behavior affects the ticagrelor's solubility. TICA form II (TICA-II) with different crystal habits were prepared by controlling the crystallization process. To systematically investigate the relationship between crystal habit and orientation of the molecules of TICA form II in the crystal lattice, single crystals were obtained, and the crystal structure is studied and reported here for first time. Morphology prediction based on BFDH (Bravais-Friedel-Donnay-Harker) theory [36,37] and face indexation [38], together with Optical Microscopy, were performed to correlate experimental and simulated crystal habits. Using X-ray powder diffraction analysis, polymorphic form conformity for different crystal habits were confirmed and preferred orientations of crystals were obtained, which were associated with the dominant crystal faces. X-Ray Photoelectron Spectroscopy (XPS) values and specific surface area were used to establish the surface chemistry. The results showed that the difference of solubility is associated with the surface anisotropy of the TICA crystal.

## 2. Experimental Section

### 2.1. Materials

Ticagrelor (TICA) form II was received from Zhejiang Ausun Pharmaceutical (Zhejiang, China). Figure 1 presents a chemical schematic of TICA. The chemical reagents used were of analytical grade.

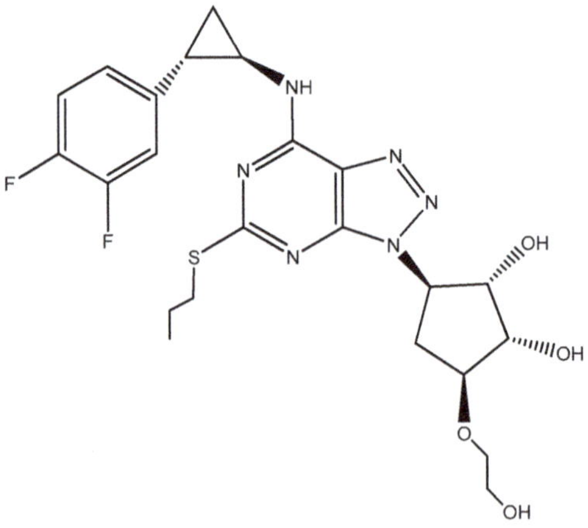

**Figure 1.** The chemical diagram of TICA.

### 2.2. Crystallization Experiments

Ticagrelor form II with different crystal habits (designated as TICA-A, TICA-B, TICA-C, TICA-D, and TICA-E) were prepared by recrystallization methods (Table S1).

TICA-A and TICA-D were crystallized from acetonitrile and butyl acetate, respectively, through rapid cooling and the mass ration of solute/solvent were 1:8 and 1:10, respectively. The solution of TICA was heated to 60 °C to ensure that no crystals remained in the solution and then underwent rapid cooling to 37 °C with stirring for 1 h. After that, the crystals were filtered and dried at 60 °C under vacuum.

TICA-B, TICA-C, and TICA-E were prepared by antisolvent methods and N-heptane used as antisolvent. The main differences are initial saturation, i.e., the mass ratio of solute and solvent (ethyl acetate). TICA was dissolved in ethyl acetate, the mass ratio (m/v) was 1:15, 1:20 and 1:10, respectively, and heated to 60 °C to dissolve completely. Stopping heating was applied and the antisolvent (n-heptane) was added to the above solutions at a 1 mL·min$^{-1}$ dropping rate under constant stirring. The antisolvent to solvent ratios for TICA-B, TICA-C and TICA-E were 1:1,1:1 and 1:1.5, respectively. The solution was then left to cool down to 25~35 °C with stirring. The obtained crystals were filtered off and dried at 60 °C under vacuum.

The single crystals of TICA form II were prepared by dissolving TICA (100 mg) in acetonitrile (18 mL) and allowing the solution to evaporate slowly. Suitable single crystals had grown after 7 days.

### 2.3. Solubility Studies

To investigate the solubility of five samples, Ultraviolet-Visible (UV) spectrophotometry was used (Thermo Scientific Evolution 300, Thermo Scientific, Waltham, MA, USA). The concentrations of TICA crystal habits were calculated by the standard curve method ($\lambda$max = 257 nm).

A beaker containing 150 mL of pH = 1.2 HCl was equilibrated at 37 °C, then approximately 150 mg of samples that had been passed through a 300 mesh sieve beforehand were added to the beaker, which was stirred at 150 rpm on a magnetic stirrer. Slurry was filtered with 0.22 μm nylon filters after 2, 4, 6, 8, 10, 15, 20, 25, 30, 40, 50, 60, 90, 120, 150, 180, 210, and 240 min. Each filtered aliquot was assayed by UV analysis at 257 nm. To ensure the accuracy of experimental data, all experiments were repeated three times.

The solubility of TICA-A, TICA-B, TICA-C, TICA-D, and TICA-E in pH = 1.2 HCl at 37 °C were measured by adding excess drug (about 150 mg) in 20 mL of pH = 1.2 HCl in a 25 mL glass bottle with screw cap. Then bottles were shaken in the magnetic stirring water bath (ALBOTE, Henan, China) at 100 rpm and kept at 37 °C (± 0.2 °C). The samples were withdrawn after 72 h, then filtered with 0.22 μm nylon filters and measured by an UV spectrometer.

### 2.4. X-Ray Powder Diffraction (PXRD)

All samples used in the PXRD experiments were sieved through 300 mesh beforehand and PXRD patterns were recorded at room temperature on a D/Max-2550PC diffractometer (Rigaku, Japan). The diffractometer was operated with monochromator Cu K$\alpha$ radiation ($\lambda$ = 1.5418 Å) at 40 kV and 250 mA. The data were recorded over a scanning range of 3~40° (2$\theta$), with an increasing step size of 0.02° (2$\theta$), and scanning speed of 3°/min.

### 2.5. Single-Crystal X-Ray Diffraction

Using a Bruker APEX-II CCD diffractometer (Bruker, Germany) with Mo K$\alpha$($\lambda$ = 0.7107 Å) radiation to collect SXRD data at −100 °C. The SAINT V8.38A [38] was used on data reduction. The absorption correction was applied with the use of semi-empirical methods of the SADABS program [39]. The crystal structure was solved by direct methods using the SHELX-S program and refined by full-matrix least-squares methods with anisotropic thermal parameters for all non-hydrogen atoms on F2 using SHELX-L [40,41]. Hydrogen atoms were placed in the position of calculation and were refined isotropically using a riding model [42]. Mercury [43] and Diamond [44] were used to draw figures.

## 2.6. Optical Microscopy

TICA crystal habits were observed for their shape and aspect ratio using a Leica DMLP polarized light microscope (Shanghai Optical Instrument Factory, China).

## 2.7. X-Ray Photoelectron Spectroscopy (XPS)

XPS were measured using a KRATOS AXIS ULTRA (DLD) (Shimadzu, Japan). The binding energy range was from 0 to 1100 eV for regions of C 1s, N 1s, O 1s, F 1s, and S 2p, with an average peak binding energy of 284.2, 397.4, 530.3, 684.8, and 160.9 eV, respectively.

## 2.8. Specific Surface Area

All samples were sieved through 300 mesh beforehand. The specific surface area was measured by the nitrogen adsorption method (Tristar II 3020 Surface Area analyzer, Micromeritics, Shanghai, China). About 100 mg samples were degassed for an hour in a vacuum environment at 80 °C to remove moisture, and then the specific surface area of the samples were calculated by the Brunauer Emmett Teller (BET) method within 0.05 to 0.2 of the relative pressure ($P/P_0$).

## 2.9. Molecular Modeling

BIOVIA Materials Studio Morphology [45] was used to predict the crystal facets of TICA from II. The TICA crystal face was first built using its CIF file. The molecular structure of acetonitrile was built using the sketching tool and geometry optimization was performed by the Forcit module using COMPASS II force field. Finally, the growth morphology of TICA crystal is given for major faces.

## 2.10. Face Indexation

The single crystal of TICA was placed onto the tip of a 0.1 mm diameter glass capillary and mounted on the Bruker APEX-II CCD diffractometer (Bruker, Germany) with CCD area detector for determining unit cell parameters and orientation matrices at −100 °C. The T-tool—the face-indexing plug-in of APEX III—was used to identify Miller indices of different faces of this crystal [38].

## 3. Results

### 3.1. Single-Crystal X-Ray Diffraction

The crystal structure of ticagrelor was studied at −100 °C and the related crystallographic data are listed in Table 1. This compound crystallizes in the P1 space group, with the asymmetric unit consisting of four ticagrelor molecules, which is similar to that of TICA form I [13]. The conformations of each TICA molecule in the asymmetric unit differ slightly and the overlay diagrams comparing different conformers of these four molecules is shown in Figure 2. It is shown that the main orientation differences are cyclopropyl-3,4 difluorophenyl, thiopropyl and hydroxyethoxy side chains, and the conformation of central groups (cyclopentane-1,2-diol-triazolopyrimidine) are almost the same. The molecular conformations in the asymmetric unit differ slightly from that of TICA form I and DMSO solvate, mainly in orientation differences of cyclopropyl-3,4 difluorophenyl and thiopropyl. Displacement ellipsoid plots showing the atomic numbering are presented in Figure 3.

Table 1. Relevant crystallographic data details for TICA-II.

| | TICA-II |
|---|---|
| Formula | $C_{23}H_{28}F_2N_6O_4S$ |
| Mr | 522.57 |
| Temperature/K | 170(2) |
| Crystal system | triclinic |
| Space group | P1 |
| a/Å | 9.8863(5) |
| b/Å | 15.7349(7) |
| c/Å | 17.6069(9) |
| α/° | 105.538(2) |
| β/° | 100.841(2) |
| γ/° | 103.091(2) |
| Volume/Å$^3$ | 2478.3(2) |
| Z | 4 |
| D/g·cm$^{-3}$ | 1.401 |
| μ/mm$^{-1}$ | 0.188 |
| F(000) | 1096.0 |
| Crystal size/mm$^3$ | 0.248 × 0.18 × 0.067 |
| Radiation | MoKα (λ = 0.71073) |
| 2θ range for data collection/° | 4.434 to 53.486 |
| Index ranges | −12 ≤ h ≤ 12, −19 ≤ k ≤ 19, −22 ≤ l ≤ 22 |
| Reflections collected | 70727 |
| Independent reflections | 20802 [Rint = 0.0541, Rsigma = 0.0563] |
| Data/restraints/parameters | 20802/6/1319 |
| Goodness-of-fit on F2 | 1.068 |
| Final R indexes [I > = 2σ (I)] | R1 = 0.0756, wR2 = 0.1948 |
| Final R indexes [all data] | R1 = 0.0886, wR2 = 0.2089 |
| Largest diff. peak/hole / e Å$^{-3}$ | 1.58/−0.33 |
| Flack parameter | 0.06(3) |
| Diffractometer | Bruker APEX-II CCD |
| Absorption correction | multi-scan |
| CCDC No. | 1953772 |

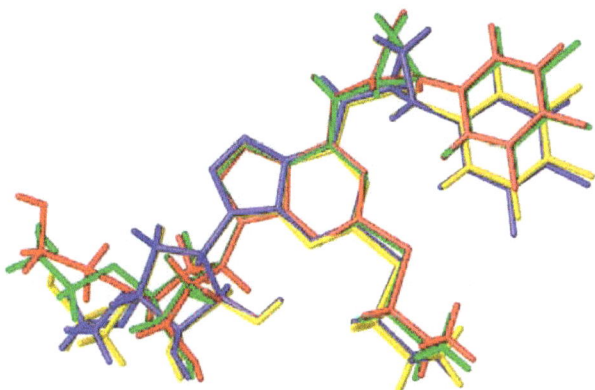

**Figure 2.** Overlay diagram for superposition of independent molecules of TICA-II crystals.

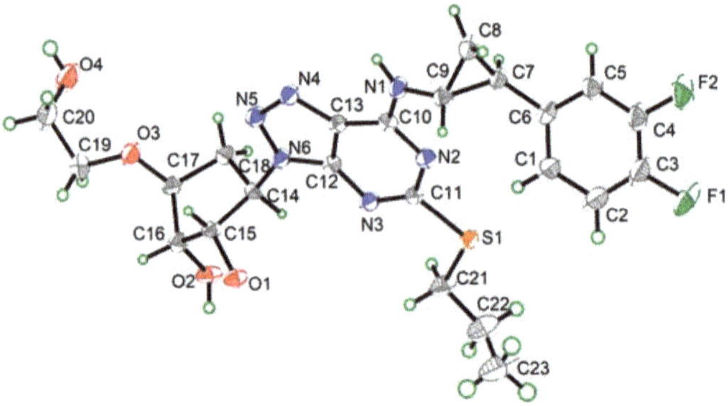

**Figure 3.** The molecule structure of TICA-II, showing displacement ellipsoids at the 50% probability level.

Obviously, the ticagrelor molecular structure contains many hydrogen-bond donors and acceptors, which justifies the existence of a wide variety of intramolecular and intermolecular hydrogen bonds. Dimeric $R_2^2(10)$ and $R_2^2(9)$ motifs between TICA molecules are generated through N-H ... N and O-H ... O intermolecular hydrogen bonds along the b-axis, forming infinite molecular chains (Figure 4), which are further stacked by hydrogen bonds between hydroxyethoxy side chains [46,47]. Thus, ring motifs between six TICA molecules are generated to form two-dimensional structures (Figure 5). TICA form I is also present in the dimeric form $R_2^2(10)$ and $R_2^2(9)$ motifs, which are formed through hydrogen bonds, but the donor and acceptor of H-bonds are different from TICA form II. H-bond data is listed in Table 2. The Hirshfeld surfaces of TICA-II is in Figure S2.

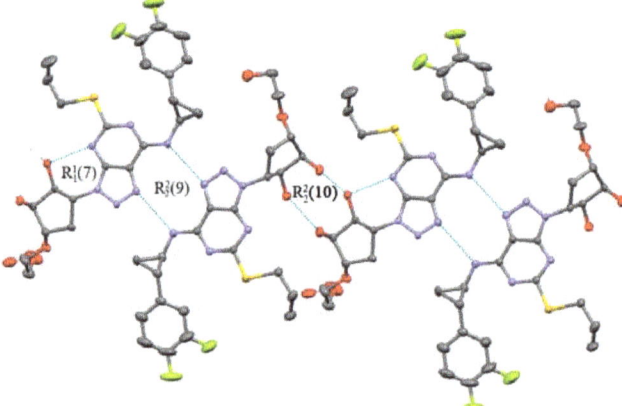

**Figure 4.** Propagation of (1) mediated by hydrogen bonds aligned along the b-axis presenting fused $R_1^1(7)$, $R_2^2(9)$ and $R_2^2(10)$ rings.

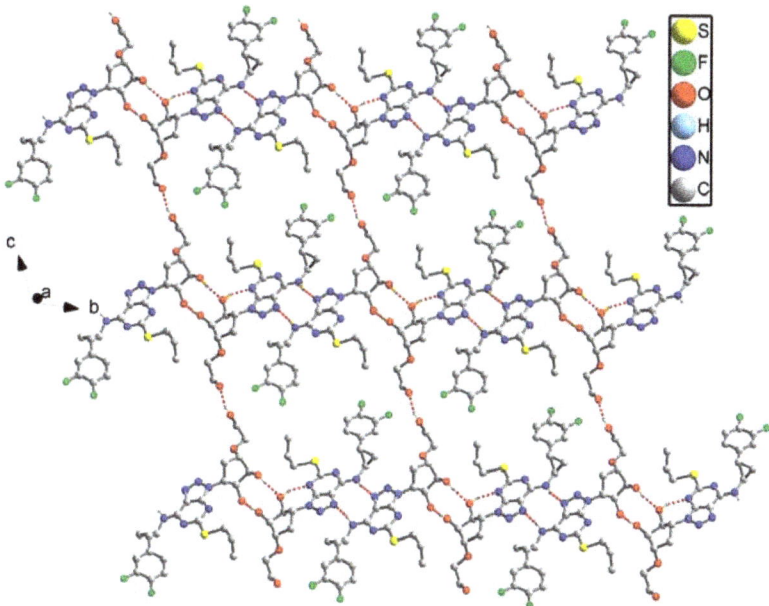

**Figure 5.** Two-dimensional hydrogen-bond networks formed by hydrogen-bond ring motifs.

**Table 2.** Hydrogen-bond of TICA-II (Å, °).

| D-H ... A | D-H | H ... A | D ... A | D-H ... A |
|---|---|---|---|---|
| O2-H2 ... O1B | 0.84 | 1.89 | 2.724(5) | 171.8 |
| O4-H4 ... O4B | 0.84 | 1.92 | 2.734(6) | 162.1 |
| N1-H1A ... N4B | 0.88 | 2.19 | 3.057(6) | 169.9 |
| O2A-H2AA ... O1C | 0.84 | 1.90 | 2.734(5) | 170.0 |
| O4A-H4A ... O4 | 0.84 | 2.01 | 2.849(6) | 174.8 |
| N1A-H1AB ... N4C | 0.88 | 2.18 | 3.045(6) | 167.4 |
| O1B-H1BA ... N3B | 0.84 | 1.86 | 2.661(5) | 158.6 |
| O2B-H2B ... O1 | 0.84 | 2.05 | 2.854(5) | 160.2 |
| O4B-H4B ... O3B | 0.84 | 2.52 | 2.856(6) | 105.4 |
| O4B-H4B ... O4C | 0.84 | 1.96 | 2.743(7) | 155.5 |
| N1B-H1BB ... N4 | 0.88 | 2.16 | 3.019(6) | 166.3 |
| O1C-H1C ... N3C | 0.84 | 1.87 | 2.665(6) | 157.7 |
| O2C-H2C ... O1A | 0.84 | 2.04 | 2.861(5) | 161.5 |
| N1C-H1CA ... N4A | 0.88 | 2.16 | 3.025(6) | 165.7 |
| O1-H1 ... O2C | 0.84(3) | 2.33(8) | 3.082(5) | 150(13) |
| O1A-H1AA ... O2B | 0.84(3) | 2.24(5) | 3.065(5) | 165(14) |
| O4C-H4C ... O3C | 0.86(3) | 2.34(11) | 2.764(7) | 111(9) |

## 3.2. Predicted Morphology of the TICA Crystal

The predicted BFDH morphology of the TICA crystal was visualized (Figure 6b). The BFDH method is a rapid method to identify the crystal morphology (hkl) most likely to form crystal habit. According to the BFDH law, the relative growth rate is inversely proportional to the d-spacing between the crystal faces. Thus, the most important morphological faces of the crystal are those with the maximum d value [37,48,49]. For TICA form II, these planes are (001) (d = 16.3 Å), (010) (d = 14.5 Å) and (01–1) (d = 13.2Å), as determined by indexation the single-crystal faces. Therefore, the predicated shape of TICA is a needle-shape. This model is in reasonable agreement with the BFDH model, when compared with the observed morphology of the crystals (Figure 6a).

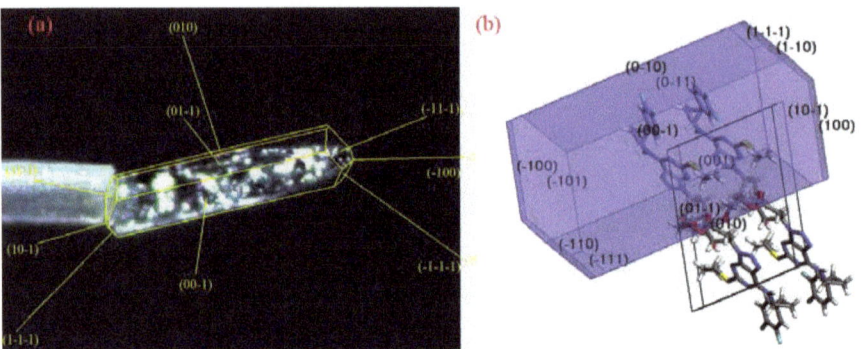

**Figure 6.** (**a**) Face indexation for TICA form II and (**b**) Morphology predictions for TICA form II by means of BFDH calculations.

### 3.3. Powder X-Ray Diffraction

The PXRD patterns of these five crystal habits (Figure 7) all had characteristic diffraction peaks at 2θ values of 5.4°, 6.7°, 13.4°, 18.3°, 22.8° and 24.8° corresponding to TICA form II reported in the literature [11], and these were also identical to those of the calculated PXRD pattern using the single-crystal diffraction results (listed at Section 3.1); however, some intensity differences were observed. The preferred orientation of the crystal results in a difference in the relative peak strength (Table 3). The most important peaks determining the growth direction of the crystal are 5.4° (d = 16.3 Å), 6.0° (d = 14.5 Å) and 6.7° (d = 13.2 Å), which correspond to the (001), (010) and (01-1) crystal faces, respectively. The diffraction peak that associated with the (010) crystal face is poorly resolved due to its low relative intensity. The intensity of the diffraction peaks corresponding to the (001), (01-1) and its relevant (02-2) crystal plane in the pattern of the above five samples were observed to vary greatly as compared with the intensities of the calculated pattern. This result suggests a varied frequency of the (001) and (01-1) planes in the five TICA crystal samples, meaning that they have different crystal habits. The DSC data is in the Figure S2.

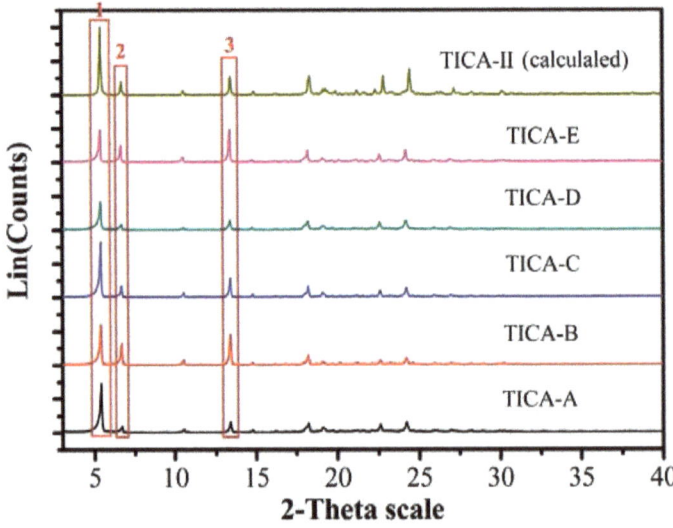

**Figure 7.** Overlay of XRD of TICA crystal habits and calculated XRD pattern of TICA form II.

Table 3. The PXRD of TICA crystal habits and calculated from single-crystal diffraction results.

| Number | 2θ (°) | Crystal Face | TICA-A | TICA-B | TICA-C | TICA-D | TICA-E | TICA-II |
|---|---|---|---|---|---|---|---|---|
| | | | \multicolumn{6}{c}{Relative Intensity ($I/I_0$)/%} | | | | | |
| 1 | 5.4 | 001 | 100 | 100 | 100 | 100 | 97.1 | 100 |
| 2 | 6.7 | 01-1 | 12.3 | 53.5 | 20.9 | 20.9 | 49.9 | 19.0 |
| 3 | 13.4 | 02-2 | 21.2 | 76.0 | 35.3 | 33.1 | 100 | 27.0 |

*3.4. Optical and Polarized Light Microscopy*

Every crystal face has its own growth rate and it is the slowest growing face which determines the growth habits of crystals. The predicted BFDH morphology of TICA-II (Figure 6b.) showed that (001), (010) and (01-1) are the main crystal planes. The crystal grows rapidly along the (010) direction, so there are few (010) crystal faces exposed. Therefore, face (001) and (01-1) become more dominant in the general shape of the crystal. From Figure 8, the crystal shapes of five samples are different, mainly in terms of their length and width. The aspect ratios of TICA-A, B, C, D, E are about 1:1~2:1, more than 10:1, mostly 5:1~8:1, mostly 1:1~3:1 and more than 10:1, respectively. When the nucleation rate is quicker than that of crystal growth, such as in the conditions of more saturation or rapid cooling (TICA-A and TICA-D), the growth rate of the (0-11) face is enhanced compared to that of face (001), resulting in a plate-like shape. However, when the crystal growth rate is quicker using the antisolvent method, enhanced growth in the (001) direction, as compared to the (0-11) face, produces needle-shaped crystals (TICA-B, C, E). Initial saturation in the crystallization process of sample TICA-C among the above three samples is highest, resulting in needle-shaped crystals with relatively good aspect ratios. The longest crystal growth time (sample TICA-E) leads to slender acicular crystals with an aspect ratio of more than 10:1 (Figure 8f).

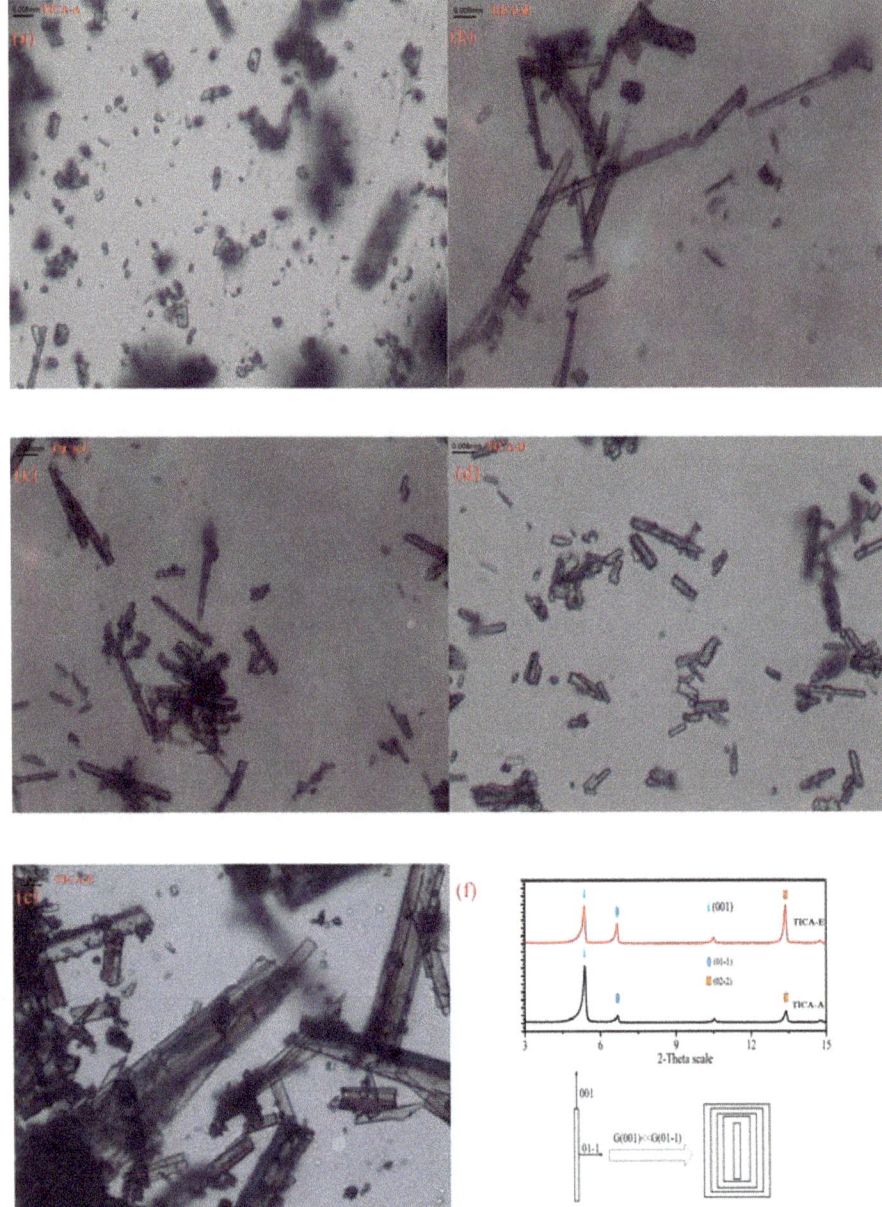

**Figure 8.** (a–e) Microscope graphs of TICA crystals and (f) Schematic diagram of the crystal growth of TICA crystal habits.

## 3.5. XPS

XPS (Table 4) showed the existence of C, O, N, F, and S on the surface of every sample. Chemical shift and peak shape for these elements were similar between the five samples, which indicate that there are no qualitative differences in TICA samples with different crystal habits. Moreover, there was no significant difference in the relative abundance of surface elements (Table 4). The (O + N + S)/(C +

F) stands for surface polarity, was 0.394, 0.372, 0.387, 0.392 and 0.375 for TICA-A, TICA-B, TICA-C, TICA-D, and TICA-E, respectively.

Table 4. The XPS data of TICA crystal habits.

| | Elemental Composition (%) | | | | | (O + N + S)/(C + F) |
|---|---|---|---|---|---|---|
| | O 1s | N 1s | S 2p | F 1s | C 1s | |
| TICA-A | 12.19 | 13.74 | 2.32 | 6.48 | 65.27 | 0.394 |
| TICA-B | 12.38 | 12.63 | 2.14 | 5.52 | 67.33 | 0.372 |
| TICA-C | 11.04 | 14.48 | 2.27 | 6.65 | 65.56 | 0.387 |
| TICA-D | 11.46 | 14.36 | 2.32 | 6.40 | 65.46 | 0.392 |
| TICA-E | 11.23 | 13.85 | 2.21 | 5.87 | 66.85 | 0.375 |

## 3.6. Solubility Study

The dissolution profiles of five samples in pH 1.2 hydrochloric acid solutions are shown in Figure 9a. The TICA-A dissolves faster than that of other four habits and the dissolution rates (in the first 30 min) are ordered as follows: TICA-A > TICA-D > TICA-B > TICA-C > TICA-E. The amounts dissolved of the above five crystal habits within 24 h are captured in Figure 9b, showing that the solubility of TICA-A and TICA-D samples whining 24 hours were higher than that of TICA-B and TICA-C samples, and the solubility of TICA-E samples was the lowest.

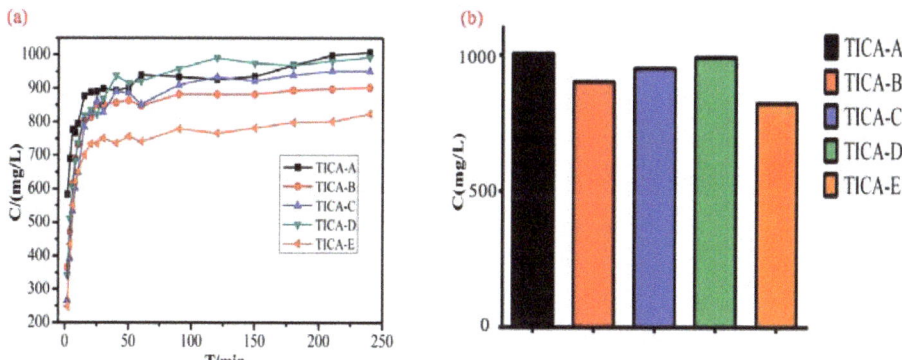

**Figure 9.** (a) Solution concentration–time profiles of TICA crystal habits in pH 1.2 hydrochloric acid solutions and (b) Amounts dissolved of TICA crystal habits in pH 1.2 hydrochloric acid solutions within 24 h.

## 3.7. Specific Surface Area

Furthermore, specific surface areas of these five samples were compared (Table 5). Because of the small specific surface area of the drug particles themselves, there will be inevitable instrument errors in the measurement. According to the test results, there was no significant difference in the specific surface area between these five samples.

Table 5. Specific Surface Area of TICA crystal Habits.

| Sample Name | Specific Surface Area (m$^2$/g) |
|---|---|
| TICA-A | 1.56 |
| TICA-B | 1.60 |
| TICA-C | 1.53 |
| TICA-D | 1.51 |
| TICA-E | 1.67 |

## 4. Discussion

The rate of crystal growth is mainly related to the rate at which units attach and remove themselves from the growing surfaces. Deposition kinetics of solute molecules on different crystal faces determine the crystal habit [22,50]. Investigations on the single-crystal structure of the TICA-II suggest that elongation is driven by the formation of hydrogen-bond networks in the crystal lattice. Thus, hydrogen bonding of TICA form II drives the crystal growth along one axis (axis b), which results in the formation of needle-shaped crystals. The (010) crystal plane demonstrates the fastest growth, and its relevant diffraction peak is almost not observable, as demonstrated in Section 3.3. Therefore, the crystal habits were mainly related to the growth rate of the (0-11) face and (001) face. Therefore, the growth rate can be changed by adjusting crystallization conditions such as solvent, temperature, degree of supersaturation, cooling rate, and stirring rate, etc. When growth rates of (001) and (0-11) were modified by controlling crystallization conditions, needle-shaped crystals with different aspect ratios and plate-like crystals of TICA form II were prepared.

The five crystal habits of TICA-II exhibited different dissolution behavior, including equilibrium solubility and dissolution rate, despite the similar powder surface area. The second law of thermodynamics governs the dissolution process. Dissolution results in the destruction of the original forces between solute molecules, forming new interactions and increasing the overall disorder. At the molecular level, dissolution can be expressed by interactions between functional groups of drug particles and solvents. The surface anisotropy of the TICA crystal face due to differential surface exposure of functional groups was known, which lead to different performance in its bioavailability. Moreover, the relative abundance of the major faces of five samples, such as (001) and (01-1), differed significantly. The relative abundance of a hydrophilic face (001) on the surface of TICA-A was higher than TICA-E. This preferred orientation leads to a higher exposure of a relatively more hydrophilic (001) facet on the surface of TICA-A and leads to its higher solubility.

The (001) crystal face has a layer of difluorophenyl and hydroxyls exposed on its surface, which makes this face more hydrophilic (Figure 10a). As for the (01-1) crystal face, the exposure of propyl groups makes it relatively more hydrophobic (Figure 10b). Therefore, it can be predicted that when the (001) crystal facet is dominant, such as in TICA-A and TICA-D, the crystals will be more hydrophilic and have better solubility. In contrast, when the dominant crystal planes are (01-1) and its relevant (02-2), such as in TICA-E, it tends to dissolve more poorly. Solubility differences for five crystal habits with different aspect ratios are possible, relative to the two crystal faces (001) and (01-1) exposed. (Figure S3.)

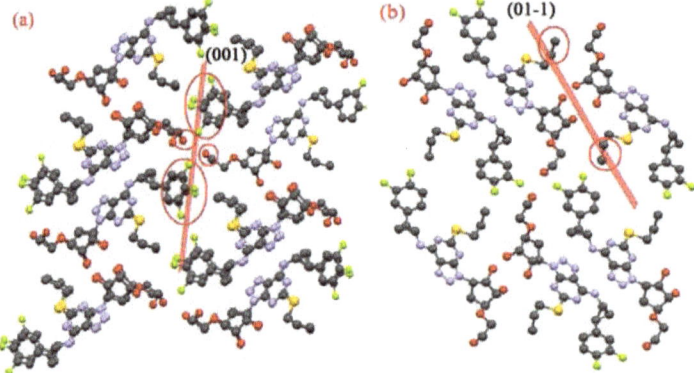

**Figure 10.** (a) Packing along (001) face and (b) (01-1) face, showing the surface chemistry.

The number of crystal faces exposed to solvent also determines the solubility of the drug [51]. The bigger the aspect ratio of the sample is, the worse the samples' solubility is. That is why the solubility of TICA-A is better than that of TICA-E. The surface polarity also affects solubility. In theory, the larger the surface polarity, the better the solubility. The specific surface area of the five crystal habits are not significantly different from each other, so the specific surface area does not affect the solubility.

## 5. Conclusions

Several crystal habits with different aspect ratios of TICA form II were prepared using different crystallization conditions. Although they have the same crystal structures, their dissolution rates are significantly different, which may be due to the difference in the surface anisotropy and abundance of exposed crystals. Modification of the surface morphology of the crystal without changing its polymorphism appears to provide a good method for enhancing the solubility of drugs, especially for BCS class IV drugs that have poor solubility.

**Supplementary Materials:** The following are available online at http://www.mdpi.com/2073-4352/9/11/556/s1, Figure S1: The DSC curve of TICA crystal habits. Table S1: Preparation method for TICA crystal habits. Figure S2: The Hirshfeld surfaces of TICA-II. Figure S3: Different screenshots of (001) face and (01-1) face using Mercury 2.3 software.

**Author Contributions:** The main contribution to this study came from Y.R. J.S. helped in the calculation. K.Y. helped in the proofreading. C.U.P. and G.C. helped in guidance. J.L. helped in single-crystal structure. J.F. and X.H. designed and supervised this work.

**Funding:** This research received no external funding.

**Acknowledgments:** The authors would like to Zhejiang Aoxiang Pharmaceutical Co., Ltd. (China, Zhejiang) for providing us with ticagrelor, also thank Zhejiang University for providing experimental facilities.

**Conflicts of Interest:** The authors states that there is no conflict of interest.

## References

1. Marzio, H.D.; Navarro, V.J. Chapter 29–Hepatotoxicity of Cardiovascular and Antidiabetic Drugs. In *Drug-Induced Liver Disease*; Academic Press: Cambridge, MA, USA, 2013; pp. 519–540.
2. Cave, B.; Rawal, A.; Ardeshna, D.; Ibebuogu, U.N.; Sai-Sudhakar, C.B.; Khouzam, R.N. Targeting ticagrelor: A novel therapy for emergency reversal. *Ann. Transl. Med.* **2019**, *7*, 115–119. [CrossRef]
3. Kubisa, M.J.; Jezewski, M.P.; Gasecka, A.; Siller-Matula, J.M.; Postula, M. Ticagrelor—toward more efficient platelet inhibition and beyond. *Ther. Clin. Risk Manag.* **2018**, *14*, 129–140. [CrossRef] [PubMed]
4. Van Giezen, J.J.J.; Berntsson, P.; Zachrisson, H.; Bjorkman, J.A. Comparison of ticagrelor and thienopyridine P2Y(12) binding characteristics and antithrombotic and bleeding effects in rat and dog models of thrombosis/hemostasis. *Thromb. Res.* **2009**, *124*, 565–571. [CrossRef] [PubMed]
5. Husted, S.; Emanuelsson, H.; Heptinstall, S.; Sandset, P.M.; Wickens, M.; Peters, G. Pharmacodynamics, pharmacokinetics, and safety of the oral reversible P2Y(12) antagonist AZD6140 with aspirin in patients with atherosclerosis: A double-blind comparison to clopidogrel with aspirin. *Eur. Heart J.* **2006**, *27*, 1038–1047. [CrossRef] [PubMed]
6. Cannon, C.P.; Husted, S.; Harrington, R.A.; Scirica, B.M.; Emanuelsson, H.; Storey, R.F.; Invest, D. Safety, tolerability, and initial efficacy of AZD6140, the first reversivle oral adenosine diphosphate receptor antagonist, compared with clopidigrel, in patients with non-ST-segment elevation acute coronary syndrome—Primary results of the DISPERSE-2 trial. *J. Am. Coll. Cardiol.* **2007**, *50*, 1844–1851. [CrossRef]
7. Chalasani, N.; Fontana, R.J.; Bonkovsky, H.L.; Watkins, P.B.; Davern, T.; Serrano, J.; Yang, H.; Rochon, J.; Drug Induced Liver Injury Network. Causes, clinical features, and outcomes from a prospective study of drug-induced liver injury in the United States. *Gastroenterology* **2008**, *135*, 1924–1934. [CrossRef]
8. Wallentin, L.; Becker, R.C.; Budaj, A.; Cannon, C.P.; Emanuelsson, H.; Held, C.; Horrow, J.; Husted, S.; James, S.; Katus, H.; et al. Ticagrelor versus Clopidogrel in Patients with Acute Coronary Syndromes. *N. Engl. J. Med.* **2009**, *361*, 1045–1057. [CrossRef]
9. Mohammad, R.A.; Goldberg, T.; Dorsch, M.P.; Cheng, J.W.M. Antiplatelet therapy after placement of a drug-eluting stent: A review of efficacy and safety studies. *Clin. Ther.* **2010**, *32*, 2265–2281. [CrossRef]
10. Reuben, A.; Koch, D.G.; Lee, W.M.; Acute Liver Failure Study Group. Drug-induced acute liver failure: Results of a U.S. multicenter, prospective study. *Hepatology* **2010**, *52*, 2065–2076. [CrossRef]

11. Martin, B. New Cristalline and Amorphous form of a Triazolo(4,5-D) pyridinine Compound. U.S. Patent WO 01/92262 A1, 6 December 2001.
12. Agarwal, V.K. Crystalline Form of Ticagrelor. U.S. Patent Application No. 16/066,425, 3 January 2017.
13. Bojarska, J.; Remko, M.; Fruzinski, A.; Maniukiewicz, W. The experimental and theoretical landscape of a new antiplatelet drug ticagrelor: Insight into supramolecular architecture directed by C-H ... F, π ... π and C-H ... π interactions. *J. Mol. Struct.* **2018**, *1154*, 290–300. [CrossRef]
14. Williams, H.D.; Trevaskis, N.L.; Charman, S.A.; Shanker, R.M.; Charman, W.N.; Pouton, C.W.; Porter, C.J.H. Strategies to address low drug solubility in discovery and development. *Pharmacol. Rev.* **2013**, *65*, 315–499. [CrossRef] [PubMed]
15. Gao, L.; Zhang, X.-R.; Yang, S.-P.; Liu, J.-J.; Chen, C.-J. Improved solubility of vortioxetine using C2-C4 straight-chain dicarboxylic acid salt hydrates. *Crystals* **2018**, *8*, 352. [CrossRef]
16. Miletic, T.; Kyriakos, K.; Graovac, A.; Ibric, S. Spray-dried voriconazole-cyclodextrin complexes: Solubility, dissolution rate and chemical stability. *Carbohydr. Polym.* **2013**, *98*, 122–131. [CrossRef] [PubMed]
17. Lu, Y.; Tang, N.; Lian, R.; Qi, J.; Wu, W. Understanding the relationship between wettability and dissolution of solid dispersion. *Int. J. Pharm.* **2014**, *465*, 25–31. [CrossRef] [PubMed]
18. Rasenack, N.; Muller, B.W. Micron-size drug particles: Common and novel micronization techniques. *Pharm. Dev. Technol.* **2004**, *9*, 1–13. [CrossRef] [PubMed]
19. Sun, C.; Grant, D.J.W. Influence of crystal shape on the tableting performance of L-lysine monohydrochloride dihydrate. *J. Pharm. Sci.* **2001**, *90*, 569–579. [CrossRef]
20. Banga, S.; Chawla, G.; Varandani, D.; Mehta, B.R.; Bansal, A.K. Modification of the crystal habit of celecoxib for improved processability. *J. Pharm. Pharmacol.* **2007**, *59*, 29–39. [CrossRef]
21. Singhal, D.; Curatolo, W. Drug polymorphism and dosage form design: A practical perspective. *Adv. Drug Deliv. Rev.* **2004**, *56*, 335–347. [CrossRef]
22. Tiwary, A.K. Modification of crystal habit and its role in dosage form performance. *Drug Dev. Ind. Pharm.* **2001**, *27*, 699–709. [CrossRef]
23. Mittal, A.; Malhotra, D.; Jain, P.; Kalia, A.; Shunmugaperumal, T. Studies on Aspirin Crystals Generated by a Modified Vapor Diffusion Method. *Aaps Pharmscitech* **2016**, *17*, 988–994. [CrossRef]
24. Serrano, D.R.; O'Connell, P.; Paluch, K.J.; Walsh, D.; Healy, A.M. Cocrystal habit engineering to improve drug dissolution and alter derived powder properties. *J. Pharm. Pharmacol.* **2016**, *68*, 665–677. [CrossRef] [PubMed]
25. Mishnev, A.; Stepanovs, D. Crystal structure explains crystal habit for the antiviral drug rimantadine hydrochloride. *Z. Fur Nat. Sect. B A J. Chem. Sci.* **2014**, *69*, 823–828. [CrossRef]
26. Maghsoodi, M. Role of solvents in improvement of dissolution rate of drugs: Crystal habit and crystal agglomeration. *Adv. Pharm. Bull.* **2015**, *5*, 13–18. [CrossRef]
27. Modi, S.R.; Dantuluri, A.K.R.; Puri, V.; Pawar, Y.B.; Nandekar, P.; Sangamwar, A.T.; Perumalla, S.R.; Sun, C.C.; Bansal, A.K. Impact of crystal habit on biopharmaceutical performance of celecoxib. *Cryst. Growth Des.* **2013**, *13*, 2824–2832. [CrossRef]
28. Di Martino, P.; Censi, R.; Malaj, L.; Capsoni, D.; Massarotti, V.; Martelli, S. Influence of solvent and crystallization method on the crystal habit of metronidazole. *Cryst. Res. Technol.* **2007**, *42*, 800–806. [CrossRef]
29. Nokhodchi, A.; Bolourtchian, N.; Farid, D. Effects of hydrophilic excipients and compression pressure on physical properties and release behavior of aspirin-tableted microcapsules. *Drug Dev. Ind. Pharm.* **1999**, *25*, 711–716. [CrossRef]
30. Ishikawa, K.; Eanes, E.D.; Tung, M.S. The effect of supersaturation on apatite crystal formation in aqueous solutions at physiologic pH and temperature. *J. Dent. Res.* **1994**, *73*, 1462–1469. [CrossRef]
31. Nokhodchi, A.; Bolourtchian, N.; Dinarvand, R. Crystal modification of phenytoin using different solvents and crystallization conditions. *Int. J. Pharm.* **2003**, *250*, 85–97. [CrossRef]
32. Shariare, M.H.; Blagden, N.; de Matas, M.; Leusen, F.J.J.; York, P. Influence of solvent on the morphology and subsequent comminution of ibuprofen crystals by air jet milling. *J. Pharm. Sci.* **2012**, *101*, 1108–1119. [CrossRef]
33. Stoica, C.; Verwer, P.; Meekes, H.; van Hoof, P.; Kaspersen, F.M.; Vlieg, E. Understanding the effect of a solvent on the crystal habit. *Cryst. Growth Des.* **2004**, *4*, 765–768. [CrossRef]

34. Umprayn, K.; Luengtummuen, A.; Kitiyadisai, C.; Pornpiputsakul, T. Modification of crystal habit of ibuprofen using the phase partition technique: Effect of Aerosil and Tween80 in binding solvent. *Drug Dev. Ind. Pharm.* **2001**, *27*, 1047–1056. [CrossRef] [PubMed]
35. Walker, E.M.; Roberts, K.J.; Maginn, S.J. A molecular dynamics study of solvent and impurity interaction on the crystal habit surfaces of epsilon-caprolactam. *Langmuir* **1998**, *14*, 5620–5630. [CrossRef]
36. Prywer, J. Explanation of some peculiarities of crystal morphology deduced from the BFDH law. *J. Cryst. Growth* **2004**, *270*, 699–710. [CrossRef]
37. Wang, Z.; Jiang, P.; Dang, L. The Morphology Prediction of Lysozyme Crystals Deduced from the BFDH Law and Attachment Energy Model Based on the Intermolecular Interaction. In Proceedings of the 2010 4th International Conference on Bioinformatics and Biomedical Engineering, Chengdu, China, 18–20 June 2010.
38. Bruker AXS announces novel APEX(TM) DUO, the most versatile system for small molecule X-ray crystallography. *Anti-Corros. Methods Mater.* **2007**, *54*, 375.
39. Sheldrick, G.M. *SADABS*; Version 2.10.; University of Gottingen: Gottingen, Germany, 2003.
40. Sheldrick, G.M. SHELXT—Integrated space-group and crystal-structure determination. *Acta Crystallogr. A Found. Adv.* **2015**, *71*, 3–8. [CrossRef]
41. Sheldrick, G.M. A short history of SHELX. *Acta Crystallogr. A Found. Adv.* **2008**, *64*, 112–122. [CrossRef]
42. Dolomanov, O.V.; Bourhis, L.J.; Gildea, R.J.; Howard, J.A.K.; Puschmann, H. OLEX2: A complete structure solution, refinement and analysis program. *J. Appl. Crystallogr.* **2009**, *42*, 339–341. [CrossRef]
43. Macrae, C.F.; Bruno, I.J.; Chisholm, J.A.; Edgington, P.R.; Mccabe, P.; Pidcock, E.; Rodriguezmonge, L.; Taylor, R.; van de Streek, D.; Wood, P.A. Mercury CSD 2.0—new features for the visualization and investigation of crystal structures. *J. Appl. Crystallogr.* **2010**, *41*, 466–470. [CrossRef]
44. Farrugia, L.J. WinGX and ORTEP for Windows: An update. *J. Appl. Crystallogr.* **2012**, *45*, 849–854. [CrossRef]
45. Li, W.D.; Zhang, M.; Li, Y.; Liu, G.X.; Li, Z.J. Effect of heat preservation time on the micro morphology and field emission properties of La-doped SiC nanowires. *Crystengcomm* **2019**, *21*, 3993–4000. [CrossRef]
46. Abourahma, H.; Cocuzza, D.S.; Melendez, J.; Urban, J.M. Pyrazinamide cocrystals and the search for polymorphs. *Crystengcomm* **2011**, *13*, 6442–6450. [CrossRef]
47. Etter, M.C. ChemInform abstract: Encoding and decoding hydrogen-bond patterns of organic compounds. *Acc. Chem. Res.* **1990**, *23*, 120–126. [CrossRef]
48. Laville, G. Elementary study on refraction in a prisma which is out of the main section. Bravais law. *J. De Phys. Et Le Radium* **1921**, *2*, 62–64. [CrossRef]
49. Donnay, J.D.H.; Harker, D. A new law of crystal morphology extending the law of bravais. *Am. Mineral.* **1937**, *22*, 446–467.
50. Hadjittofis, E.; Isbell, M.A.; Karde, V.; Varghese, S.; Ghoroi, C.; Heng, J.Y.Y. Influences of Crystal Anisotropy in Pharmaceutical Process Development. *Pharm. Res.* **2018**, *35*, 100. [CrossRef]
51. Bukovec, P.; Benkic, P.; Smrkolj, M.; Vrecer, F. Effect of crystal habit on the dissolution behaviour of simvastatin crystals and its relationship to crystallization solvent properties. *Pharmazie* **2016**, *71*, 263–268. [CrossRef]

© 2019 by the authors. Licensee MDPI, Basel, Switzerland. This article is an open access article distributed under the terms and conditions of the Creative Commons Attribution (CC BY) license (http://creativecommons.org/licenses/by/4.0/).

*Article*

# Synthesis, Crystal Structure, and Solubility Analysis of a Famotidine Cocrystal

Yan Zhang [1,2,*], Zhao Yang [2], Shuaihua Zhang [3] and Xingtong Zhou [2]

1. School of Medicine and Pharmacy, Ocean University of China, Qingdao 266003, China
2. Qingdao Institute for Food and Drug Control, Qingdao 266071, China
3. School of Pharmacy, Qingdao University, Qingdao 266071, China
* Correspondence: yanzhang139@139.com; Tel.: +86-532-5875-9178

Received: 15 May 2019; Accepted: 10 July 2019; Published: 15 July 2019

**Abstract:** A novel cocrystal of the potent $H_2$ receptor antagonist famotidine (FMT) was synthesized with malonic acid (MAL) to enhance its solubility. The cocrystal structure was characterized by X-ray single crystal diffraction, and the asymmetry unit contains one FMT and one MAL connected via intermolecular hydrogen bonds. The crystal structure is monoclinic with a P21/n space group and unit cell parameters a = 7.0748 (3) Å, b = 26.6502 (9) Å, c = 9.9823 (4) Å, α = 90, β = 104.2228 (12), γ = 90, V = 1824.42 (12) Å$^3$, and Z = 4. The cocrystal had unique thermal, spectroscopic, and powder X-ray diffraction (PXRD) properties that differed from FMT. The solubility of the famotidine-malonic acid cocrystal (FMT-MAL) was 4.2-fold higher than FMT; the FAM-MAL had no change in FMT stability at high temperature, high humidity, or with illumination.

**Keywords:** cocrystal; famotidine; malonic acid; crystal structure; solubility

## 1. Introduction

Drugs with low water solubility usually show dissolution-limited absorption and low bioavailability [1]. Recent estimates suggest that approximately 40% of currently marketed drugs and up to 75% of compounds currently under development are poorly water soluble; thus, enhancing the aqueous solubility of poorly water-soluble active pharmaceutical ingredients (APIs) is a key challenge for pharmaceutical scientists [2]. Salts are historically the first choice for overcoming poor solubility and dissolution rate problems in APIs. However, cocrystals have recently emerged as a subject of intense research [3].

In the past decade, cocrystal technology emerged as an advanced approach to enhance the aqueous solubility of poorly water-soluble drugs via crystal engineering principles without changing their chemical structure [4–12]. A pharmaceutical cocrystal is defined as a multicomponent molecular complex combining an API and coformer(s) through non-covalent interactions (e.g., hydrogen bonding, van der Waals forces, π-stacking, and electrostatic interactions) in a definite stoichiometric ratio [13,14]. Distinctions between salts and cocrystals can be made based on whether a proton transfer has occurred from an acid to a base [15]. However, the determination of whether a salt or a cocrystal has formed can be difficult. Crystal structure determination often does not afford accurate proton positions, and other techniques are, therefore, often necessary [16].

Famotidine(3-[({2-[(Hydrazonomethyl)amino]thiazol-4-yl}methyl)thio]-*N*-sulfamoylpropionamidine) (FMT, pKa = 7.06) (Figure 1) is a potent $H_2$ receptor antagonist commonly used for gastroesophageal reflux disease [17]. It is insoluble in cold water, and this poor aqueous solubility may contribute to its low and variable oral bioavailability [18]. FMT exhibits two major crystalline polymorphs (forms A and B), where B is the metastable form [19]. Several studies have suggested that cocrystal solubility can correlate strongly with coformer solubility [20], thus, we describe here a water-soluble organic acid—Malonic acid

(MAL; $pKa_1$ = 2.8, $pKa_2$ = 5.7) (Figure 1). This was selected as cocrystal coformer (CCF) to synthesize a novel cocrystal with FMT. Single crystal X-ray diffraction, Fourier transform infrared spectroscopy (FT-IR), and conductivity experiments were performed to identify that it is a cocrystal instead of a salt. Furthermore, the cocrystal of FMT with MAL (FMT-MAL) showed a 4.2-fold increase in FMT solubility without changing its stability.

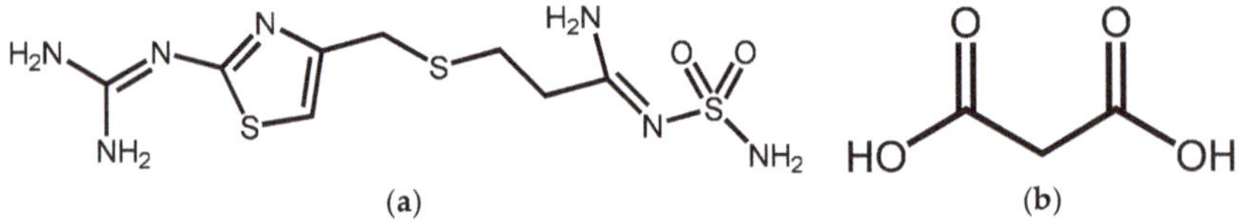

Figure 1. Structures of (a) FMT and (b) MAL.

## 2. Experimental Section

### 2.1. General

FMT (drug substance, form B) was supplied by Qingdao Liteng Chemical Medical Research Co., Ltd (Qingdao, China). A reference standard of FMT (100305–201304, 99.5% purity) was purchased from National Institutes for Food and Drug Control. High-pressure liquid chromatography (HPLC) grade methanol and acetonitrile were purchased from Merck (Darmstadt, Germany). MAL and the other chemicals were purchased from Sinopharm Chemical Reagent Co., Ltd. (Shanghai, China). Powder diffractograms were recorded on a Bruker D8 Focus with a Cu-Kα radiation (1.54060 Å). A simulated powder X-ray diffraction (PXRD) pattern was calculated from the refined single crystal structures of FMT-MAL by Mercury 4.0.0. A Spectrum65 FT-IR spectrometer (Perkin Elmer) was employed in the KBr diffuse-reflectance mode (sample concentration was 2 mg in 20 mg of KBr) for collecting the IR spectra of the sample. Differential scanning calorimetry (DSC) used a DSC6000 (Perkin Elmer, Waltham, Massachusetts, USA)). Thermogravimetric analysis (TGA) was performed using a TGA8000 instrument (Perkin Elmer, Waltham, MA, USA).

### 2.2. Synthesis of FMT-MAL Cocrystal

FMT (340 mg, 1 mmol) was dissolved in a mixture of methanol (7 mL) and dimethylformamide (3 mL) and heated and stirred at 70 °C for 1 h. MAL (104 mg, 1 mmol) was added and stirred at 70 °C for 2 h. The resulting solution was filtered and allowed to slowly evaporate at room temperature for 2 days. Single crystals were carefully selected under a microscope and kept in mineral oil.

### 2.3. Single-Crystal X-ray Diffraction Analysis

X-ray reflections were collected on a D8 Quest diffractometer (Bruker, Rheinstetten, Germany) equipped with a PHOTON II CPAD detector using Co radiation (0.71073 Å) at room temperature (296 K). Data were corrected via a ω scan, and the absorption effects were studied using the multi-scan method. The structure was solved with the ShelXT [21] structure solution program using direct methods and refined with the XL [22] refinement package using Least Squares minimization. Anisotropical thermal factors were assigned to all of the non-hydrogen atoms. The positions of the hydrogen atoms were generated geometrically.

### 2.4. Conductivity

Conductivity was measured using a Sioinlab740 conductivity meter (Mettler Toledo, Greifensee, Switzerland). FMT-MAL (52 mg) and a mixture of FMT (40 mg) and MAL (12 mg) were both dissolved in 35 mL water.

## 2.5. High-Pressure Liquid Chromatography (HPLC)

A solution of each sample (for solubility and stability experiments) was analyzed via a1260 HPLC system (Agilent Technologies, Santa Clara, California, USA) equipped with a diode array detector (set at 270 nm). The chromatographic separation used an Agilent ZORBAX SB-C18 5 µm column (4.6 mm × 250 mm) at 35 °C. The mobile phase consisted of sodium acetate buffer (13.6 mg mL$^{-1}$, pH = 6.0 ± 0.1) and acetonitrile (93:7 in volume) at a flow rate of 1.5 mL min$^{-1}$.

## 2.6. Solubility Determination

The solid samples for solubility studies were sieved (150 µm ± 6.6 µm diameter mesh sieve) to obtain uniform particle size. The solubility was determined by suspending 500 mg of the solid samples in 100 mL of water at 25 °C and 100 rpm in 708-DS dissolution apparatus (Agilent Technologies, Santa Clara, CA, USA) with a paddle method. Samples (1 mL) were collected at 5, 10, 20, 30, 45, 60, 90,120, 240, 360, 480, 600, 720, 840, 960, 1080, 1200, 1320, and 1440 min, and then filtered through a 0.45-µm nylon filter and assayed for drug content by HPLC.

## 2.7. Stability Determination

We next tested the stability at high relative moisture content, high temperature, and high illumination conditions. The FMT and FMT-MAL were stored in a SHH-SDF stability chamber (Yongsheng, Chongqing, China) at 60 °C, 25 °C/92.5% relative humidity (RH), and 4500 lux conditions for 10 days. Samples were collected at 0, 5, and 10 days, and analyzed by HPLC to assay for chemical purity.

## 3. Results and Discussion

### 3.1. Crystal Strcture

The cocrystal showed unique peaks in PXRD that differed from its raw components, indicating that a new cocrystal was formed. The PXRD pattern of FMT-MAL showed new peaks at 2θ values of 14.2°, 16.2°, 18.5°, 21.6°, 22.0°, 27.8°, and 29.8°, which were not present in FMT and MAL (Figure 2). A further comparison between the simulated PXRD pattern and the experimental PXRD pattern of FMT-MAL showed a definitive match indicating the homogeneity and purity of FMT-MAL crystalline phase.

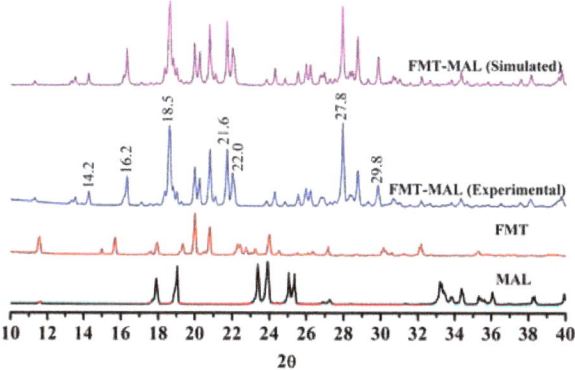

**Figure 2.** PXRD patterns of FMT-MAL (simulated), FMT-MAL (experimental), FMT, and MAL.

The FMT-MAL was a colorless block, unlike the colorless needle of FMT. X-ray structural analyses revealed that FMT-MAL crystallizes in a non-standard space group ($P2_1/n$). Crystal data for FMT-MAL are summarized in Table 1. Selected bond lengths and angles for FMT-MAL are listed in Table 2. The asymmetric units of FMT-MAL consist of one molecule of FMT and one molecule of MAL (Figure 3a).

There are three kinds of intermolecular H-bonds within the molecular units (Figure 3b). The first and second ones share the same donor atom (N1). The first one is N1–H1A···O2 (pick dotted line, $d_{N···O}$ = 2.827 (1) Å; ∠NHO = 145°). The second one is N1–H2B···O3 (green dotted line, $d_{N···O}$ = 2.796 (1) Å; ∠NHO = 171.00°). Here, the O3 from the carbonyl group of MAL is the acted as an acceptor. The third H-bond originates from N2–H2A···O6 (turquoise dotted line, $d_{N···O}$ = 2.837 (1) Å; ∠NHO = 167°), in which the O6 from the carbonyl group of MAL acted as an acceptor. The fourth hydrogen bond originates from N5–H5B···O1 (black dotted line, $d_{N···O}$ = 2.860 (1) Å; ∠NHO = 157°). Here, the O1 from sulfanilamide group acted as an acceptor. The hydrogen bonds for FMT-MAL are listed in Table 3. The hydrogen bonding interactions in the structure of FMT-MAL allowed the FMT and MAL to connect in a molecular way and expand the networks to three dimensions (Figure 3c). The crystallographic data, in CIF format, was deposited with the Cambridge Crystallographic Data Centre, CCDC 1576773.

Table 1. Crystal data and structure refinements.

| Compound | FMT-MAL |
|---|---|
| Formula | $C_{11}H_{19}N_7O_6S_3$ |
| Formula Weight | 441.51 |
| Temperature (K) | 296 (2) |
| Crystal System | Monoclinic |
| Space Group | $P2_1/n$ |
| a (Å) | 7.0748 (3) |
| b (Å) | 26.6502 (9) |
| c (Å) | 9.9823 (4) |
| $\alpha$ ° | 90 |
| $\beta$ ° | 104.2228 (12) |
| $\gamma$ ° | 90 |
| Volume (Å$^3$) | 1824.42 (12) |
| Z | 4 |
| Density calculated (g·cm$^{-3}$) | 1.607 |
| F (000) | 920 |
| Absorption coefficient (mm$^{-1}$) | 0.453 |
| Reflection collected | 25641 |
| Unique Reflection | 8367 |
| $R_{int}$ | 4189 |
| $R_1{}^a$, $wR_2{}^b$ [I > 2σ(I)] | 0.0544/0.1477 |
| $R_1$, $wR_2$ (all data) | 0.0691/0.1586 |
| Goodness-of-fit on $F^2$ | 1.049 |
| $\Delta\rho_{max}$, $\Delta\rho_{min}$, (e Å$^{-3}$) | 0.617, −0.559 |

Note: [a] $R_1=\Sigma||F_o|-|F_c||/\Sigma|F_o|$, [b] $wR_2=[\Sigma w(F_o{}^2-F_c{}^2)^2/\Sigma w(F_o{}^2)^2]^{1/2}$.

Table 2. Selected Bond Lengths (Å), Angles (°), and Torsion angles (°).

| FMT-MAL | | | |
|---|---|---|---|
| S (1)-C (3) | 1.719 (3) | S (1)-C (2) | 1.723 (2) |
| S (2)-C (6) | 1.805 (3) | S (2)-C (5) | 1.814 (3) |
| S (3)-O (1) | 1.432 (2) | S (3)-N (6) | 1.613 (2) |
| S (3)-O (2) | 1.438 (2) | S (3)-N (7) | 1.609 (3) |
| C (9)-O (3) | 1.222 (4) | C (11)-O (5) | 1.300 (4) |
| C (9)-O (4) | 1.278 (4) | C (11)-O (6) | 1.215 (4) |
| N (3)-C (1) | 1.355 (3) | N (3)-C (2) | 1.391 (3) |
| N (4)-C (2) | 1.295 (3) | N (4)-C (4) | 1.391 (3) |
| O (1)-S (3)-N (7) | 109.30 (15) | C (1)-N (3)-C (2) | 125.2 (2) |
| O (2)-S (3)-N (7) | 105.97 (13) | C (18)-N (2)-C (21) | 108.02 (14) |
| O (1)-S (3)-N (6) | 105.50 (12) | N (2)-C (1)-N (1) | 122.0 (2) |
| O (2)-S (3)-N (6) | 113.49 (12) | N (2)-C (1)-N (3) | 121.0 (2) |
| N (7)-S (3)-N (6) | 106.09 (13) | N (1)-C (1)-N (3) | 117.0 (2) |
| O (1)-S (3)-N (6)-C (8) | | −168.8 (2) | |
| O (2)-S (3)-N (6)-C (8) | | −40.7 (3) | |
| N (7)-S (3)-N (6)-C (8) | | 75.2 (2) | |

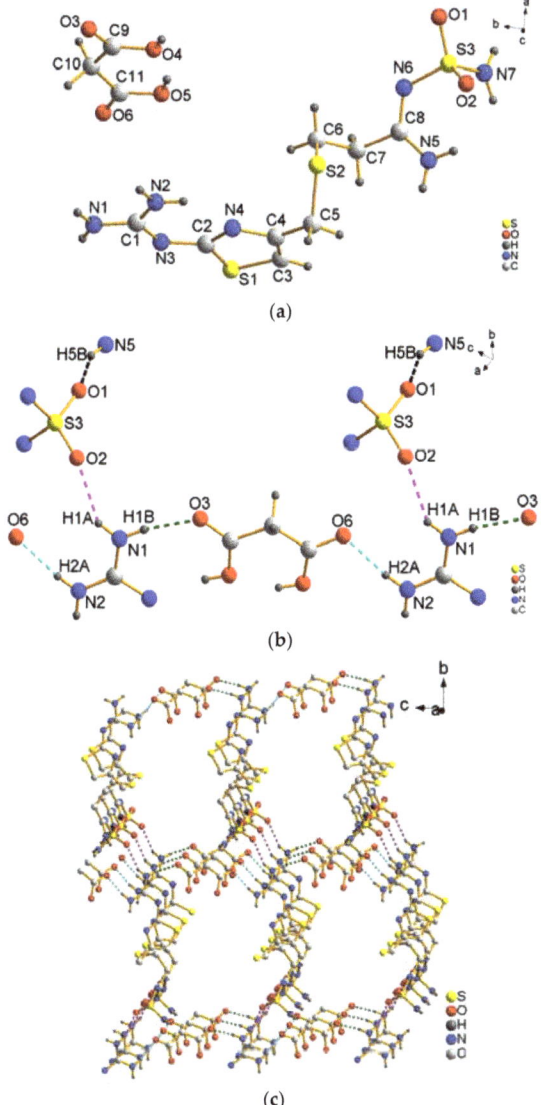

**Figure 3.** (a) View of the asymmetric unit of FMT-MAL, (b) hydrogen bonds in FMT-MAL, and (c) view of the 3D supramolecular structure of FMT-MAL.

**Table 3.** Hydrogen-bond geometries and interactions for FMT-MAL.

| Hydrogen Bond | Distance [a], Å | Distance [b], Å | Angle [c], ° |
|---|---|---|---|
| N1–H1A···O2 | 2.18 | 2.927 (1) | 145 |
| N1–H1B···O3 | 1.94 | 2.796 (1) | 171 |
| N2–H2A···O6 | 1.99 | 2.837 (1) | 167 |
| N5–H5B···O1 | 2.05 | 2.860 (1) | 157 |

Note: [a] Distance between donor and acceptor; [b] distance between hydrogen and acceptor; [c] angle of acceptor–hydrogen–donor.

During the structure analysis, N3 of guanidine was found to show approximately 0.65e/Å$^3$ residual electron, which is located approximately 0.93 Å positions from N3 toward O4 of malonic acid. The distance of C1–N3 is 1.355 (3) Å. This is much shorter than the distance of C–N in [C = NH$^+$] (1.371 (1) Å) versus the reported structure of FMT-HCl [23]. This suggests that the N3 atom from C = N group has not been protonated. The carbonyl groups of malonic acid have shorter bond lengths (C9–O3 = 1.222 (4) Å and C11–O6 = 1.215 (4) Å) and C–OH groups of malonic acid have larger bond lengths (C9–O4 = 1.278 (4) Å and C11–O5 = 1.300 (4) Å). Thus, malonic acid in the cocrystal exists as a neutral moiety (no proton was transfer occurred), which is evident from the carboxylic acid bond length [24–26]. Therefore, the residual electron around N3 was rejected as a hydrogen atom, and one hydrogen atom was calculated on O4 to make the malonic acid a neutral molecule and not the anion form. The flat conformation of the neutral malonic acid molecule, which has two OH groups (O4 and O5) on the same side of the molecule, is reported in a few structures, such as CUVDON, CUVDON01, EPERIP, UFETOR, UFEVUZ, and URMALN (ConQuest Version 2.0.0), thus it is not rare. Further study of the sulfanilamide group showed that the distances of S3–N6 and S3–N7 are 1.613 (2) Å and 1.609 (3) Å, respectively. The N7–S3–N6 bridging angle is 106.09 (13)°. The distances and angles are comparable to reported FMT crystal structure, suggesting that the NH$_2$ from the sulfanilamide group has not been protonated [27]. In addition, we performed conductivity experiments. The conductivity studies showed that the solution of FMT-MAL had a lower conductivity (190.3 μS cm$^{-1}$) than the mixture of FMT and MAL at a molar ratio of 1:1 (244.7 μS cm$^{-1}$). The lower conductivity suggests that FMT-MAL may still be connected together via weak interactions in the water. Thus, the results of the conductivity experiment suggest that FMT-MAL is a cocrystal rather than a salt at room temperature.

### 3.2. Thermal Properties

DSC analysis revealed that FMT-MAL had a unique melting point at 172.5 °C, which was higher than the melting points of both FMT (163.6 °C) and MAL (135.3 °C) (Figure 4a). It is common that the cocrystal shows an intermediate melting point between the API and conformer. However, a study showed that within the survey, 50 cocrystalline samples were analyzed, and 3 out of 50 (6%) had melting points higher than either the API or conformer. Higher thermal stability of the cocrystal could be associated to crystal packing [28]. The TGA curves show that FMT-MAL had an onset decomposition temperature of 180.0 °C. There was no significant change in the weight before an endothermic phenomenon occurred. This confirmed its non-solvated character and high purity (Figure 4b).

### 3.3. Fourier Transform Infrared Spectroscopy

Infrared spectroscopy is a powerful tool for detecting cocrystal formation because the vibrational changes serve as probes for intermolecular interactions in solid materials [28]. The famotidine showed peaks at 3505, 3399, and 3376 cm$^{-1}$, corresponding to the –NH stretch. There are decreases in the –N–H stretching frequency of famotidine at 3389, 3342, and 3246 cm$^{-1}$ in the cocrystal. The peak of malonic acid at 1700 cm$^{-1}$ corresponds to the –C=O stretch and moved to high-frequency (1721 cm$^{-1}$) in the cocrystal (Figure 5). This suggests that the –N–H group of famotidine and the –C=O group of malonic acid participants have some interactions. The results were coincident with the single-crystal X-ray diffraction data. Furthermore, a rapid means of establishing the formation of a salt or a cocrystal of FMT with MAL is by detection of shifts in ν (C = O) for the MAL. An indication of cocrystal formation is the occurrence of a shift of C = O group of malonic acid to higher energies (from 1700 cm$^{-1}$ to 1721 cm$^{-1}$) [3].

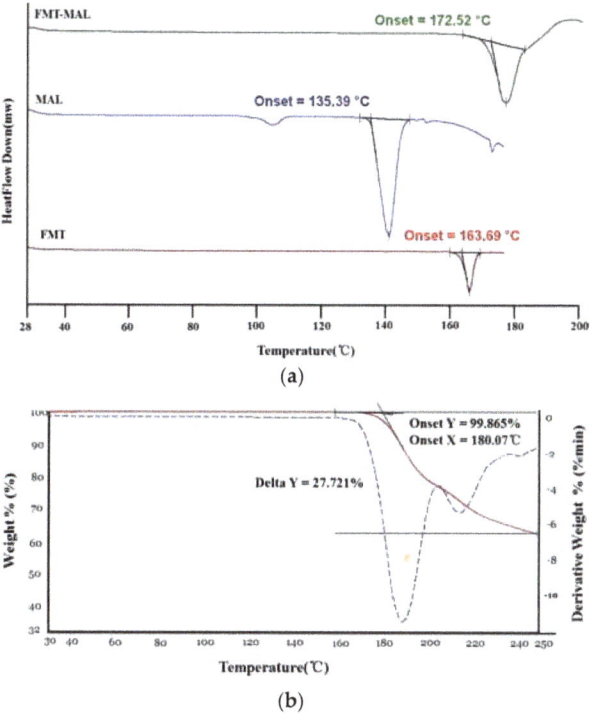

Figure 4. (a) DSC and (b) TGA thermograms of FMT-MAL.

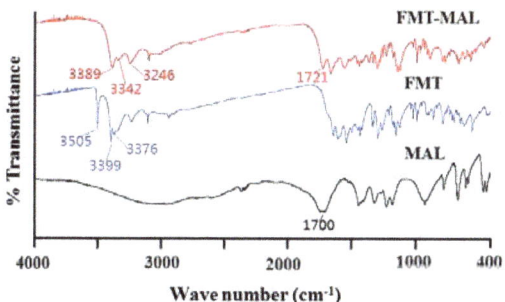

Figure 5. IR spectra of the FMT-MAL, FMT, and MAL.

## 3.4. Aqueous Solubility

The maximum famotidine concentration was determined to be 0.96 mg mL$^{-1}$ after 8 h. The solubility enhancement was demonstrated via the FMT-MAL cocrystal—the maximum concentration of FMT was 4.06 mg mL$^{-1}$ after 30 min. Thus, the FMT-MAL showed a 4.2-fold increase versus parent FMT (Figure 6). The cocrystal was stable for more than 24 h, as confirmed by analyzing undissolved material using PXRD (Figure 7).

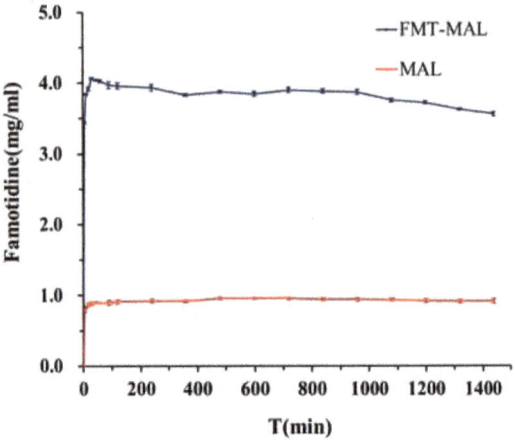

**Figure 6.** Solubility curves of FMT and FMT-MAL.

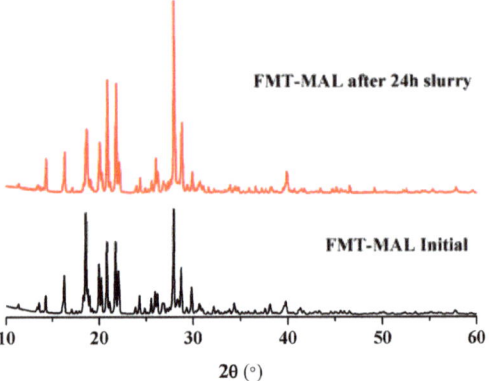

**Figure 7.** PXRD patterns of FMT-MAL and product after 24 h in a slurry of water.

### 3.5. Stability

The stability studies of FMT and FMT-MAL at 60 °C, 25 °C/92.5% RH, and 4500 lux conditions indicated that the cocrystal did not change the stability of FMT (Table 4).

**Table 4.** Stability data at 60 °C, 25 °C/92.5% RH, and 4500 lux conditions.

| FMT | Inspection Item | 0 Day | 5 Day | 10 Day |
|---|---|---|---|---|
| 4500 lux | Content (%) | 99.8 | 99.4 | 98.9 |
| 60 °C | Content (%) | 99.8 | 99.2 | 98.2 |
| 25 °C, 92.5% RH | Increasing Weight (%) | 0 | 0.005 | 0.008 |
|  | Content (%) | 99.8 | 98.8 | 97.6 |
| **FMT-MAL** | **Inspection Item** | **0 Day** | **5 Day** | **10 Day** |
| 4500 lux | Content (%) | 99.5 | 99.3 | 99.0 |
| 60 °C | Content (%) | 99.5 | 98.2 | 97.5 |
| 25 °C, 92.5% RH | Increasing Weight (%) | 0 | 0.007 | 0.010 |
|  | Content (%) | 99.5 | 98.6 | 98.5 |

## 4. Conclusions

In summary, we describe a novel cocrystal of FMT obtained by co-crystallization with MAL. This improves its solubility. The cocrystal was characterized by single-crystal X-ray diffraction. The crystal structure was monoclinic with a P21/n space group; the asymmetry unit contained a FMT and a MAL connected via intermolecular hydrogen bonds between the amide of famotidine and the carboxy of malonic acid. The FMT-MAL had unique thermal, spectroscopic, and PXRD properties that differed from FMT. The FMT-MAL improved the aqueous solubility of famotidine. There was a 4.2-fold increase in FMT solubility with no impact on stability. This new cocrystal can improve the bioavailability of FMT but additional trials are needed to confirm it. These results offer further insight into the co-crystallization in terms of both supramolecular chemistry and solubility modification.

**Author Contributions:** Data curation, Y.Z., Z.Y., and X.Z.; methodology, Y.Z. and S.Z.; writing—original draft, Y.Z.; writing—review and editing, Y.Z. and Z.Y.

**Funding:** This research received no external funding.

**Conflicts of Interest:** The authors declare no conflict of interest.

## References

1. Kawabata, Y.; Wada, K.; Nakatani, M.; Yamadaa, S.; Onouea, S. Formulation design for poorly water-soluble drugs based on biopharmaceutics classification system: Basic approaches and practical applications. *Int. J. Pharm.* **2011**, *420*, 1–10. [CrossRef] [PubMed]
2. Williams, H.D.; Trevaskis, N.L.; Charman, S.A.; Shanker, R.M.; Charman, W.N.; Pouton, C.W.; Porter, C.J.H. Strategies to address low drug solubility in discovery and development. *Pharmacol. Rev.* **2013**, *65*, 315–499. [CrossRef] [PubMed]
3. Mittapalli, S.; Mannava, M.K.C.; Khandavilli, U.B.R.; Allu, S.; Nangia, A. Soluble salts and cocrystals of clotrimazole. *Cryst. Growth Des.* **2015**, *15*, 2493–2504. [CrossRef]
4. Park, B.; Yoon, W.; Yun, J.; Ban, E.; Yun, H.; Kim, A. Emodin-nicotinamide (1:2) cocrystal identified by thermal screening to improve emodin solubility. *Int. J. Pharm.* **2019**, *557*, 26–35. [CrossRef] [PubMed]
5. Gao, Y.; Zu, H.; Zhang, J. Enhanced dissolution and stability of adefovir dipivoxil by cocrystal formation. *J. Pharm. Pharmacol.* **2011**, *63*, 483–490. [CrossRef] [PubMed]
6. Mcnamara, D.P.; Childs, S.L.; Giordano, J.; Iarriccio, A.; Cassidy, J.; Shet, M.S.; Mannion, R.; O'Donnell, E.; Park, A. Use of a glutaric acid cocrystal to improve oral bioavailability of a low solubility API. *Pharm. Res.* **2006**, *23*, 1888–1897. [CrossRef] [PubMed]
7. Zeng, Q.Z.; Ouyang, J.; Zhang, S.; Zhang, L. Structural characterization and dissolution profile of mycophenolic acid cocrystals. *Eur. J. Pharm. Sci.* **2017**, *102*, 140–146. [CrossRef] [PubMed]
8. Stavropoulos, K.; Johnston, S.C.; Zhang, Y.G.; Rao, B.G.; Hurrey, M.; Hurter, P.; Topp, E.M.; Kadiyala, I. Cocrystalline solids of telaprevir with enhanced oral absorption. *J. Pharm. Sci.* **2015**, *104*, 3343–3350. [CrossRef]
9. Wyche, T.P.; Alvarenga, R.F.R.; Piotrowski, J.S.; Duster, M.N.; Warrack, S.R.; Cornilescu, G.; De Wolfe, T.J.; Hou, Y.; Braun, D.R.; Ellis, G.A.; et al. Chemical genomics, structure elucidation, and in vivo studies of the marine-derived anticlostridial ecteinamycin. *ACS Chem. Biol.* **2017**, *12*, 2287–2295. [CrossRef]
10. Sowa, M.; Slepokura, K.; Matczakjon, E.; Sowa, M.; Ślepokura, K.; Matczak-Jon, E.J.C. Improving solubility of fisetin by cocrystallization. *Cryst. Eng. Comm.* **2014**, *16*, 10592–10601. [CrossRef]
11. Drozd, K.V.; Manin, A.N.; Churakov, A.V.; Perlovich, G.L. Drug-drug cocrystals of antituberculous 4-aminosalicylic acid: Screening, crystal structures, thermochemical and solubility studies. *Eur. J. Pharm. Sci.* **2017**, *99*, 228–239. [CrossRef] [PubMed]
12. Chadha, R.; Bhandari, S.; Haneef, J.; Khullar, S.; Mandal, S. Cocrystals of telmisartan: Characterization, structure elucidation, in vivo and toxicity studies. *Cryst. Eng. Comm.* **2014**, *16*, 8375–8389. [CrossRef]
13. Qiao, N.; Li, M.; Schlindwein, W.; Malek, N.; Davies, A.; Trappitt, G. Pharmaceutical cocrystals: An overview. *Int. J. Pharm.* **2011**, *419*, 1–11. [CrossRef] [PubMed]
14. Miroshnyk, I.; Mirza, S.; Sandler, N. Pharmaceutical co-crystals-an opportunity for drug product enhancement. *Expert Opin. Drug Del.* **2009**, *6*, 333–341. [CrossRef] [PubMed]

15. Childs, S.L.; Stahly, G.P.; Park, A. The salt-cocrystal continuum: The influence of crystal structure on ionization state. *Mol. Pharm.* **2007**, *4*, 323–338. [CrossRef] [PubMed]
16. Cerreia Vioglio, P.; Chierotti, M.R.; Gobetto, R. Pharmaceutical aspects of salt and cocrystal forms of APIs and characterization challenges. *Adv. Drug Deliv. Rev.* **2017**, *117*, 86–110. [CrossRef] [PubMed]
17. Rad, T.S.; Khataee, A.; Kayan, B.; Kalderis, D.; Akay, S. Synthesis of pumice-TiO$_2$ nanoflakes for sonocatalytic degradation of famotidine. *J. Clean Prod.* **2018**, *202*, 853–862. [CrossRef]
18. Mady, F.M.; Abou-Taleb, A.E.; Khaled, K.A.; Yamasaki, K.; Iohara, D.; Ishiguro, T.; Hirayama, F.; Kaneto, U.; Otagiri, M. Enhancement of the aqueous solubility and masking the bitter taste of famotidine using drug/SBE-β-CyD/povidone K30 complexation approach. *J. Pharm. Sci.* **2010**, *99*, 4285–4294. [CrossRef]
19. Német, Z.; Hegedűs, B.; Szántay, C.; Sztatisz, J.; Pokol, G. Pressurization effects on the polymorphic forms of famotidine. *Thermochim. Acta* **2005**, *430*, 35–41. [CrossRef]
20. Karashima, M.; Kimoto, K.; Yamamoto, K.; Kojima, T.; Ikeda, Y. A novel solubilization technique for poorly soluble drugs through the integration of nanocrystal and cocrystal technologies. *Eur. J. Pharm. Biopharm.* **2016**, *107*, 142–150. [CrossRef]
21. Sheldrick, G.M. SHELXT–Integrated space-group and crystal-structure determination. *Acta Crystallogr. A Found. Adv.* **2015**, *71*, 3–8. [CrossRef] [PubMed]
22. Sheldrick, G.M. Crystal structure refinement with SHELXL. *Acta Crystallogr. C Struct. Chem.* **2015**, *71*, 3–8. [CrossRef] [PubMed]
23. Ishida, T.; In, Y.; Doi, M.; Inoue, M.; Yanagisawa, I. Structural study of histamine H2-receptor antagonists. Five 3-[2-(diaminomethyleneamino)-4-thiazolylmethylthio] propionamidine and -amide derivatives. *Acta Crystallogr. B Struct. Sci.* **1989**, *45*, 505–512. [CrossRef] [PubMed]
24. Lemmerer, A.; Bernstein, J.; Kahlenberg, V. One-pot covalent and supramolecular synthesis of pharmaceutical co-crystals using the API isoniazid: A potential supramolecular reagent. *CrystEngComm* **2010**, *12*, 2856–2864. [CrossRef]
25. Noonan, T.J.; Chibale, K.; Cheuka, P.M.; Bourne, S.A.; Caira, M.R. Cocrystal and salt forms of an imidazopyridazine antimalarial drug lead. *J. Pharm. Sci.* **2019**, *108*, 2349–2357. [CrossRef]
26. Shimpi, M.R.; Alhayali, A.; Cavanagh, K.L.; Rodríguez-Hornedo, N.; Velaga, S.P. Tadalafil–malonic acid cocrystal: Physicochemical characterization pH-solubility, and supersaturation studies. *Cryst. Growth Des.* **2018**, *18*, 4378–4387. [CrossRef]
27. Yanagisawa, I.; Hirata, Y.; Ishii, Y. Studies on histamine H$_2$ receptor antagonists. 2. Synthesis and pharmacological activities of N-sulfamoyl and N-sulfonyl amidine derivatives. *J. Med. Chem.* **1987**, *30*, 1787–1793. [CrossRef]
28. Schultheiss, N.; Newman, A. Pharmaceutical cocrystals and their physicochemical properties. *Cryst. Growth Des.* **2009**, *9*, 2950–2967. [CrossRef]

© 2019 by the authors. Licensee MDPI, Basel, Switzerland. This article is an open access article distributed under the terms and conditions of the Creative Commons Attribution (CC BY) license (http://creativecommons.org/licenses/by/4.0/).

Article

# Comparative Evaluation of the Photostability of Carbamazepine Polymorphs and Cocrystals

Reiko Yutani [1,*], Ryotaro Haku [2], Reiko Teraoka [1,3,*], Chisato Tode [4], Tatsuo Koide [5], Shuji Kitagawa [6], Toshiyasu Sakane [1] and Toshiro Fukami [2,*]

1. Laboratory of Pharmaceutical Technology, Kobe Pharmaceutical University, 4-19-1 Motoyamakita-machi, Higashinada-ku, Kobe 658-8558, Japan
2. Department of Molecular Pharmaceutics, Meiji Pharmaceutical University, 2-522-1 Noshio, Kiyose, Tokyo 204-8588, Japan
3. Department of Pharmaceutical Technology, College of Pharmaceutical Sciences, Himeji Dokkyo University, 7-2-1 Kamiohno, Himeji, Hyogo 670-8524, Japan
4. Instrumental Analysis Center, Kobe Pharmaceutical University, 4-19-1 Motoyamakita-machi, Higashinada-ku, Kobe 658-8558, Japan
5. Division of Drugs, National Institute of Health Sciences, 3-25-26 Tonomachi, Kawasaki-ku, Kawasaki, Kanagawa 210-9501, Japan
6. Kobe Pharmaceutical University, 4-19-1 Motoyamakita-machi, Higashinada-ku, Kobe 658-8558, Japan
* Correspondence: r-yutani@kobepharma-u.ac.jp (R.Y.); teraoka@gm.himeji-du.ac.jp (R.T.); fukami@my-pharm.ac.jp (T.F.); Tel.: +81-78-441-7532 (R.Y.); +81-79-223-6833 (R.T.); +81-42-495-8936 (T.F.)

Received: 24 September 2019; Accepted: 22 October 2019; Published: 24 October 2019

**Abstract:** Carbamazepine (CBZ), a widely used antiepileptic, is known to be sensitive to light. The aim of this study was to evaluate the photostabilities of three cocrystals of CBZ (CBZ–succinic acid (SUC), CBZ–saccharin (SAC) form I, and CBZ–SAC form II) illuminated with a $D_{65}$ fluorescent lamp compared with those of the conventional solid forms: CBZ polymorphs (forms I, II, and III). The order of discoloration determined using a colorimetric measurement was almost consistent with that of the degradation rates estimated using Fourier-transform infrared reflection–absorption spectroscopy, and these parameters of CBZ polymorphs increased in the order of form III, form I, and form II. CBZ–SUC and CBZ–SAC form I significantly suppressed the discoloration and degradation of CBZ compared with the raw CBZ, while CBZ–SAC form II facilitated the discoloration and degradation of CBZ. These results were supported by the results from the low-frequency Raman spectroscopy. The molecular mobility estimated using solid-state nuclear magnetic resonance $^1$H spin–lattice relaxation time strongly correlated with the degradation rate constant, indicating that molecular mobility significantly decreased following the formation of CBZ–SUC and CBZ–SAC form I and resulted in higher photostability. Overall, CBZ–SUC and CBZ–SAC form I are photostable forms and cocrystallization was proven to be an effective approach to improving the photostability of a photolabile drug.

**Keywords:** carbamazepine; photostability; polymorphs; cocrystal; succinic acid; saccharin

---

## 1. Introduction

The efficacy, safety, and quality of pharmaceuticals are adversely affected by light exposure. Most drugs are exposed to light during production, storage, distribution, handling, and use [1,2]. Light potentially alters physicochemical properties of the active pharmaceutical ingredients (APIs) and affects the stability of the final products [1–3]. Light can provoke the photodecomposition of a drug, which may cause not only a loss of potency of the drug but also the production of a highly toxic compound. Thus, a better understanding of the nature and extent of photodecomposition, the mechanisms of the degradation, and the wavelength causing instability is needed to stabilize the product [2,3].

Carbamazepine (CBZ), a widely used antiepileptic, is known to be a photosensitive drug and the form issued by the manufacturer has been reported to exhibit a color change after 3 days of illumination with a xenon lamp. Form diversity of CBZ is also well known, and many reports have examined its polymorphs [4,5], hydrate and solvate forms [5,6], and the relationship between its solid form and physicochemical properties [7,8].

Recently, cocrystals have received considerable attention as a new class of API solids. A cocrystal is defined as a multiple crystalline molecular complex coexisting through noncovalent interactions such as hydrogen bonding, π–π stacking interactions, and van der Waals forces [9,10]. Cocrystals can alter the physicochemical properties of the API without modifying its intrinsic structure [10]. In addition, the cocrystallization technique has been applied to a wide range of drugs because it does not require proton transfer between components, unlike salt formation [11]. CBZ has been shown to form cocrystals with various cocrystal formers (coformers), including dicarboxylic acid and saccharin (SAC) [12]. Favorable physicochemical properties of these cocrystals have been reported previously. Ullah et al. reported the advantages of a cocrystal of CBZ with succinic acid (SUC) (CBZ–SUC) in terms of aqueous solubility, in vitro drug release, and bioavailability [13]. According to Hickey et al., a cocrystal of CBZ with SAC (CBZ–SAC form I) potentially represents a viable alternative to the conventional anhydrous polymorph due to its physical and chemical stability and oral bioavailability [14]. As shown in a study by Porter III et al., a polymorph exists in a cocrystal of CBZ with SAC (CBZ–SAC form II) [15]. Although many reports have analyzed CBZ cocrystals, no studies have investigated their photostabilities. In addition, fewer studies have focused on the photosensitivity of cocrystals compared with those concentrated on the conventional solid forms [16–19].

Thus, in the present study, we prepared two polymorphs (CBZ form I and form II) and three cocrystals of CBZ (CBZ–SUC, CBZ–SAC form I, and CBZ–SAC form II) and compared their photostabilities with CBZ form III, which is known to be a comparatively stable form of CBZ. Photostability was assessed using colorimetric measurements, low-frequency Raman spectroscopy, Fourier-transform infrared reflection–absorption spectroscopy (FT-IR RAS) and solid-state nuclear magnetic resonance (SSNMR) $^1$H spin–lattice relaxation time ($T_1$).

## 2. Materials and Methods

### 2.1. Materials

CBZ (purity: ≥97.0%) was purchased from Fujifilm Wako Pure Chemical Corporation (Osaka, Japan). SUC (purity: ≥99.5%) and SAC (purity: ≥98.0%) were purchased from Nacalai Tesque, Inc. (Kyoto, Japan). The chemical structures of CBZ, SUC, and SAC are shown in Figure 1. All the other reagents were of analytical grade or reagent grade.

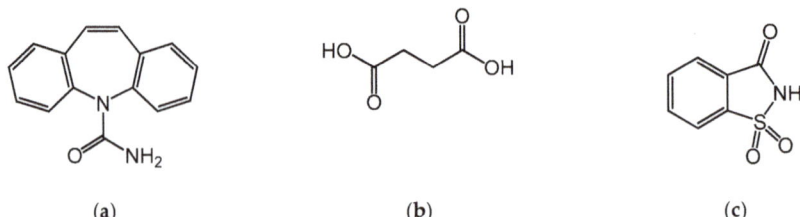

**Figure 1.** Chemical structures of (**a**) carbamazepine (CBZ), (**b**) succinic acid (SUC), and (**c**) saccharin (SAC).

### 2.2. Preparation of CBZ Polymorphs

CBZ polymorphs (forms I and II) were prepared according to a previously reported method [20].

CBZ form I: CBZ was dissolved in a 50% ethanol solution at 70 °C, stored at room temperature and dried in vacuo for 3 h, obtaining CBZ·dihydrate. Then, CBZ form I was obtained by drying CBZ·dihydrate in vacuo at 115 °C for 6 h.

CBZ form II: CBZ was dissolved in chloroform, after which diethyl ether was added and stirred at room temperature. The precipitated crystals were then filtered and dried in vacuo in a desiccator containing $P_2O_5$ at room temperature for 3 h, obtaining CBZ form II.

CBZ form III: The commercial product was used without further purification.

### 2.3. Preparation of CBZ Cocrystals

Three cocrystals were prepared according to previously reported methods [13–15].

For the preparation of CBZ–SUC, 1.00 g of CBZ and 0.25 g of SUC (molar ratio: 2:1) were dissolved in 30 mL of ethanol at 70 °C and stored at room temperature. The precipitated crystals were then filtered and dried in vacuo at room temperature.

For the preparation of CBZ–SAC form I, 1.04 g of CBZ and 0.81 g of SAC (molar ratio: 1:1) were dissolved in 20 mL of ethanol, stored overnight at 50 °C and then maintained at 30 °C. The precipitated crystals were filtered and dried in vacuo at room temperature.

For the preparation of CBZ–SAC form II, 1.04 g of CBZ and 0.81 g of SAC (molar ratio: 1:1) were dissolved in 20 mL of ethanol at 70 °C and stored overnight at room temperature. The precipitated crystals were filtered and dried in vacuo at room temperature.

As a control, physical mixtures (PMs) of CBZ form III and SUC or SAC were prepared by weighing CBZ and each coformer in an equivalent molar ratio to cocrystal and mixing them with a vortex mixer.

### 2.4. Powder X-Ray Diffractometry (PXRD)

PXRD patterns of CBZ polymorphs, coformers, PMs, and cocrystals were recorded using an X-ray diffractometer (RINT-Ultima, Rigaku Co., Tokyo, Japan) with CuK$\alpha$ radiation generated at 36 kV and 20 mA at room temperature. Data were collected within a diffraction angle range of 5°–40° (2θ) at a step size of 0.02° and a scanning speed of 2°/min.

### 2.5. Thermal Analysis

Differential scanning calorimetry (DSC) analyses were performed using a DSC 3500 Sirius instrument (NETZSCH Japan K.K., Yokohama, Japan). Two milligrams of each sample were weighed in an aluminum pan and heated from 30 to 240 °C at a rate of 10.0 °C/min.

### 2.6. Colorimetric Measurement

One hundred milligrams of CBZ polymorph, PMs, and cocrystal powders were compressed to prepare tablets with 10 mm in diameter using a compression/tension testing machine (TG-50kN, MinebeaMitsumi Inc., Tokyo, Japan) equipped with flat-faced punches and a cylindrical die. The compressed tablets were stored in a light-irradiation tester (Light-Tron LT-120, Nagano Science Co. Ltd., Osaka, Japan) equipped with a $D_{65}$ fluorescent lamp. The illuminance was set to 3500 lx. The irradiation tests were conducted at 25 °C.

The surface color of the tablet was measured with a color reader (CR-13, Konica Minolta Japan, INC., Tokyo, Japan) after the designated irradiation times. The color difference ($\Delta E^*_{ab}$) before and after irradiation was calculated using Equation (1) to evaluate the degree of discoloration. All values were reported as the averages of three measurements:

$$\Delta E^*_{ab} = \sqrt{\left(L^*_t - L^*_0\right)^2 + \left(a^*_t - a^*_0\right)^2 + \left(b^*_t - b^*_0\right)^2}, \quad (1)$$

where $\Delta E^*_{ab}$ is the color difference, $L^*$ is the brightness, $a^*$ is the red/green coordinate, and $b^*$ is the yellow/blue coordinate.

The apparent discoloration rate constant (*k*) was calculated using Equation (2) [21]:

$$\log \Delta E^*_{ab} = \frac{1}{1-n}\log t + \frac{1}{1-n}\log[(1-n)k]. \quad (2)$$

## 2.7. Low-Frequency Raman Spectroscopy

The CBZ form III, CBZ–SUC, or CBZ–SAC form II powder was ground using a mortal and a pestle, and then passed through a No. 200 sieve (nominal aperture size: 75 µm). The sieved powder was mixed with microcrystalline cellulose at a weight ratio of 1:9 (the sieved powder/ microcrystalline cellulose) for 5 min. Two hundred milligrams of the resulting mixture were weighed and compressed to prepare tablets with a diameter of 8 mm.

The sample tablets were irradiated with a $D_{65}$ fluorescent lamp for 343 h (over 1.2 million lx·h) as described above (Section 2.6). Qualitative analyses of changes in the surface of the tablets after irradiation were conducted using the Raman WorkStation (Kaiser Optical Systems Inc., MI, USA) equipped with a low-frequency XLF-CLM module (ONDAX Inc., CA, USA). The excitation wavelength was set to 976 nm. The standard spectrum of each component was obtained with an exposure time of 10 s in three scans. Samples with an area of 500 µm × 500 µm were scanned with a spectral range from −200 to 200 $cm^{-1}$, a step size of 16.7 µm, and a resolution of 4.0 $cm^{-1}$. The data were analyzed using ISys 5.0 software (Malvern Instruments Ltd., Worcestershire, UK). The spectral analysis was performed in a range of 5–200 $cm^{-1}$. The spectrum of each component was preprocessed using the standard normal variate (SNV), and images were generated based on the standard spectrum of each API using the partial least squares (PLS) analysis.

## 2.8. FT-IR RAS

Five hundred milligrams of CBZ polymorph, PMs, and cocrystal powders were compressed to prepare tablets with a diameter of 20 mm. The prepared tablets were fixed on a glass plate and irradiated with a $D_{65}$ fluorescent lamp as described above (Section 2.6). FT-IR RAS spectra were obtained using a Frontier FT-IR system (PerkinElmer Japan Co., Ltd.). The spectra were modified by the Kramers–Kronig transform and baseline correction was performed. The remaining percentage was calculated from the intensity ratio of the C=O stretching vibration (at 1675–1685 $cm^{-1}$) of CBZ before and after irradiation [22]. The photodegradation rate constant was also calculated from the apparent first-order plots.

## 2.9. SSNMR $^1H$ Spin–Lattice Relaxation Time ($T_1$)

Spin–lattice relaxation times ($T_1$) of CBZ polymorphs and cocrystals were measured using a Varian VXR $^1H$ SSNMR spectrometer (Varian Inc., Palo Alto, CA, USA) at 500 MHz. Cross-polarization and magic angle spinning (CP-MAS) methods were applied with a spinning rate of 20 kHz, a contact time of 200 µs, and a recycle delay of 600 s, using a zirconia sample tube with a diameter of 3.2 mm. Saturation recovery times were assayed from 0.1 to 1500 s.

## 3. Results

### 3.1. Characterization of CBZ Polymorphs and Cocrystals

The PXRD patterns of CBZ polymorphs, two coformers, PMs, and cocrystals are shown in Figure S1 (Supplementary Materials). The PXRD patterns of CBZ form I, form II, and form III were consistent with the patterns reported by Grzesiak et al. (Figure S1a–c, Supplementary Materials) [4]. The PXRD patterns of CBZ–SUC, CBZ–SAC form I, and CBZ–SAC form II, which were also consistent with the patterns described in previous reports, were different from those of CBZ form III, the corresponding coformers, and PMs (Figure S1c–j, Supplementary Materials) [13–15].

The DSC profiles of CBZ polymorphs, two coformers, PMs, and cocrystals are shown in Figure S2 (Supplementary Materials). The DSC profiles of each polymorph were approximately consistent with the profiles described in a previous report (Figure S2a–c, Supplementary Materials) [4]. The DSC profiles of CBZ–SUC, CBZ–SAC form I, and CBZ–SAC form II exhibited the characteristic endothermic peaks at 191.7, 177.9, and 174.5°C, respectively, which differed from those of CBZ form III, the corresponding coformers, and PMs (Figure S2c–j, Supplementary Materials). These results confirmed the formation of cocrystals.

## 3.2. Changes in the Surface Color of Tablets

Changes in the appearance of sample tablets were evaluated by calculating the difference in color before and after irradiation with a $D_{65}$ fluorescent lamp and the apparent discoloration rate constant ($k$). Time courses for discoloration of CBZ polymorphs, PMs, and cocrystals as well as their apparent discoloration rate constants are shown in Figure 2 and Table 1, respectively. The surface color of tablets gradually turned from white to pale yellow-white following light exposure, and $\Delta E^*_{ab}$ increased as the irradiation time increased. The discoloration rate constants of CBZ polymorphs increased in the order of form III, form I, and form II, and the surface color of the tablet of form II turned yellow. The surface color of the tablets of PMs also changed, similar to that of the raw CBZ. On the other hand, the discoloration rate constant of CBZ cocrystals differed. The discoloration rate constants of CBZ–SUC and CBZ–SAC form I were significantly lower than those of the raw CBZ and the corresponding PMs, that is, changes in appearance were significantly suppressed. Meanwhile, the discoloration rate constant of CBZ–SAC form II was significantly higher and the surface color of the tablet more quickly turned yellow.

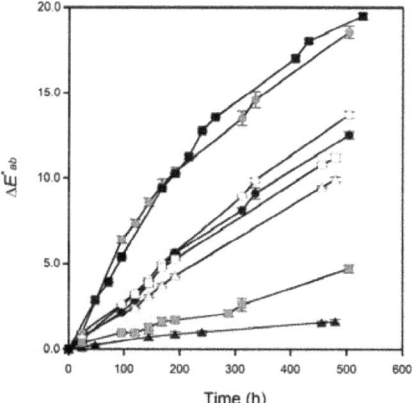

**Figure 2.** Time courses of the discoloration of CBZ polymorphs, physical mixtures (PMs), and cocrystals. The symbols represent changes in $\Delta E^*_{ab}$ of CBZ form I (○), form II (●), form III (●), CBZ–SUC PM (△), CBZ–SUC (▲), CBZ–SAC PM (□), CBZ–SAC form I (■), and CBZ–SAC form II (■). Data are presented as the means ± SD of 3 values.

**Table 1.** Apparent discoloration rate constants of CBZ polymorphs, PMs, and cocrystals.

| Sample | Apparent Discoloration Rate Constant ($h^{-1}$) |
|---|---|
| CBZ form I | 0.032 |
| CBZ form II | 0.054 |
| CBZ form III | 0.026 |
| CBZ–SUC PM | 0.024 |
| CBZ–SUC | 0.004 |
| CBZ–SAC PM | 0.026 |
| CBZ–SAC form I | 0.008 |
| CBZ–SAC form II | 0.064 |

## 3.3. Evaluation of Photodegradation on the Surface of Tablets Using Low-Frequency Raman Spectroscopy

Photodegradation of the surface of sample tablets of CBZ form III, CBZ–SUC, and CBZ–SAC form II under a $D_{65}$ fluorescent lamp was evaluated using low-frequency Raman spectroscopy. Low-frequency Raman spectroscopy is a simple and highly sensitive analytical technique for identifying the crystal form [23,24]. Standard spectra of CBZ form III, CBZ–SUC, CBZ–SAC form II, and microcrystalline cellulose are shown in Figure S3 (Supplementary Materials). Blue-to-red domains in each image represent the low-to-high density of each API. For CBZ form III and CBZ–SUC (Figure 3a,b), the Raman images and representative spectra of each domain before and after 343 h of irradiation were not substantially different, indicating that these crystal forms were comparatively photostable. In contrast, the blue domain of CBZ–SAC form II increased and the characteristic peaks of the cocrystal in the domain almost disappeared after 343 h of irradiation (Figure 3c), indicating that part of CBZ–SAC form II photodegraded after irradiation.

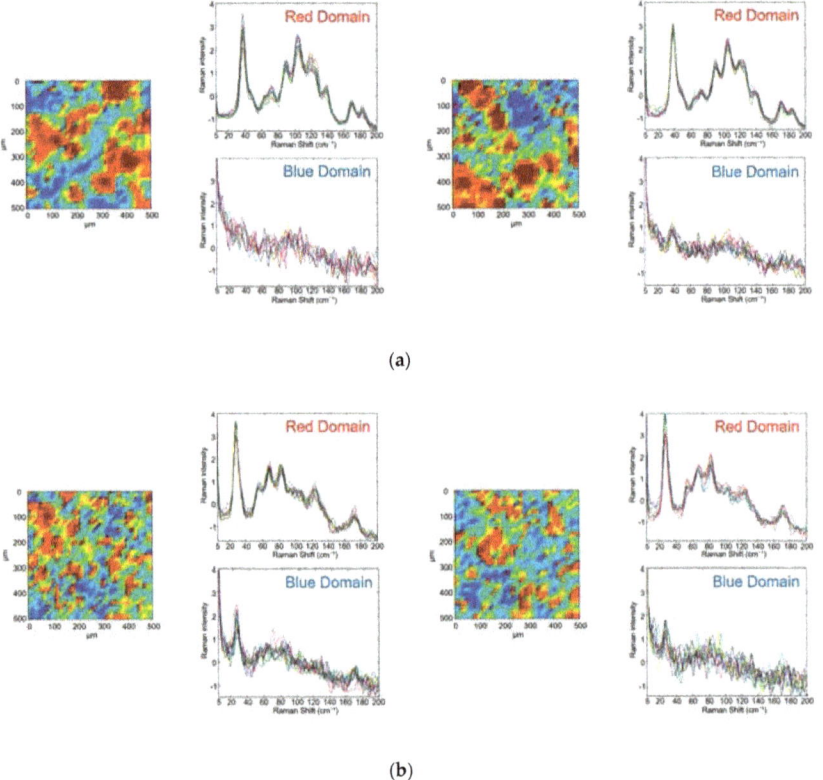

Figure 3. Cont.

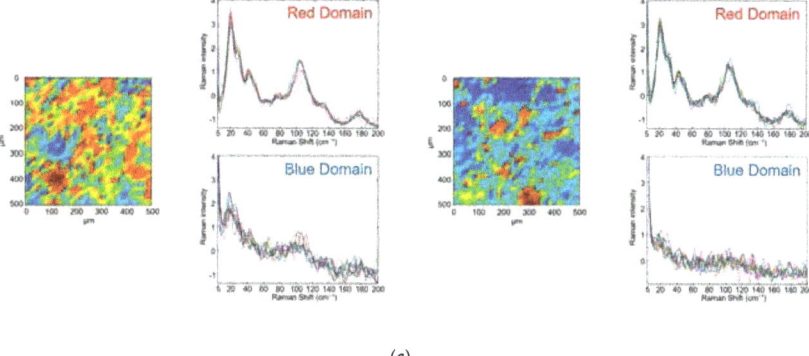

(c)

**Figure 3.** Low-frequency Raman images and representative spectra of the red domain (upper panels) and blue domain (lower panels) before irradiation (left panels) and after 343 h of irradiation (right panels): (**a**) CBZ form III, (**b**) CBZ–SUC, and (**c**) CBZ–SAC form II. The Raman spectra of each domain were obtained ten times.

### 3.4. Evaluation of Photodegradation on the Surface of Tablets Using FT-IR RAS

An evaluation of photodegradation of the surface of the sample tablets of CBZ polymorphs, PMs, and cocrystals under a $D_{65}$ fluorescent lamp was also performed using FT-IR RAS. Since the photodegradation of pharmaceuticals in tablet form is a topochemical reaction, this method is suitable for evaluating the decomposition ratio of a drug on the surface of tablets in a quantitative and reproducible manner [22,25,26]. Figure 4 shows a semilogarithmic plot of the degradation profiles of CBZ polymorphs, PMs, and cocrystals. Negative linear relationships were observed between the logarithm of the remaining percentage and irradiation time for each sample, indicating that the degradation of the drug on the surface followed first-order kinetics. Calculated degradation rate constants are shown in Table 2. The degradation rate constants of CBZ polymorphs increased in the order of form III, form I, and form II, consistent with the order of the discoloration rate constants. Degradation rate constants for CBZ–SUC and CBZ–SAC form I were significantly lower than those for the other forms, indicating that the photostability was improved by the formation of these cocrystals. In contrast, the degradation rate constant of CBZ–SAC form II was significantly increased, indicating that CBZ–SAC form II was a photolabile crystal form.

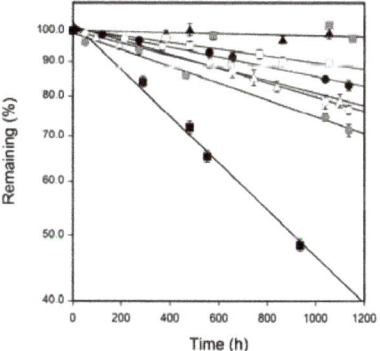

**Figure 4.** The apparent first-order plots for the photodegradation of CBZ polymorphs, PMs, and cocrystals. The symbols represent changes in the remaining percentages of CBZ form I (○), form II (●), form III (●), CBZ–SUC PM (△), CBZ–SUC (▲), CBZ–SAC PM (□), CBZ–SAC form I (■), and CBZ–SAC form II (■). Data are presented as the means ± SD of 3 values.

**Table 2.** Degradation rate constants for CBZ polymorphs, PMs, and cocrystals.

| Sample | Degradation Rate Constant ($h^{-1}$) |
|---|---|
| CBZ form I | $5.53 \times 10^{-4}$ |
| CBZ form II | $6.51 \times 10^{-4}$ |
| CBZ form III | $3.78 \times 10^{-4}$ |
| CBZ–SUC PM | $4.86 \times 10^{-4}$ |
| CBZ–SUC | $3.79 \times 10^{-5}$ |
| CBZ–SAC PM | $2.63 \times 10^{-4}$ |
| CBZ–SAC form I | $4.23 \times 10^{-5}$ |
| CBZ–SAC form II | $1.82 \times 10^{-3}$ |

### 3.5. Investigation of the Relationship between Molecular Mobility and Photostability

Spin–lattice relaxation times ($T_1$) of CBZ polymorphs and cocrystals were measured using $^1$H SSNMR to investigate the relationship between molecular mobility and photostability. The spin–lattice relaxation time reflects the degree of order and mobility, i.e., the dynamics, in the molecular system [27,28]. Figure 5 shows the relationship between $^1$H $T_1$ values and degradation rate constants. The $^1$H $T_1$ value of each polymorph was in the order of form II, form I, and form III, that is, the more photostable crystal form had a larger $^1$H $T_1$ value. A negative linear correlation was observed between the logarithm of $^1$H $T_1$ and degradation rate constant. Thus, the molecular mobility was significantly decreased by the formation of CBZ–SUC and CBZ–SAC form I, leading to higher photostability.

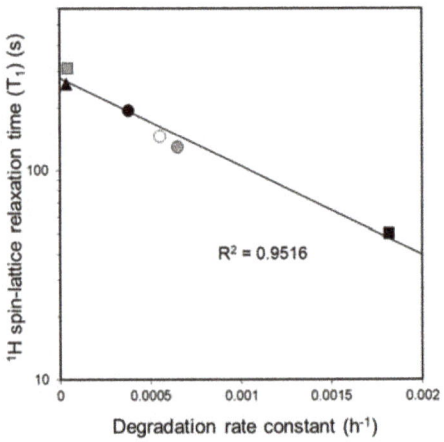

**Figure 5.** The relationship between the $^1$H spin–lattice relaxation time ($T_1$) and degradation rate constant of CBZ polymorphs and cocrystals: CBZ form I (○), form II (◉), form III (●), CBZ–SUC (▲), CBZ–SAC form I (■), and CBZ–SAC form II (■).

## 4. Discussion

The cocrystallization technique improves the physicochemical properties of an API without modifying its chemical structure. In the present study, we prepared three cocrystals, including two polymorphs, of CBZ and investigated their photostabilities compared with that of the CBZ polymorphs. CBZ–SUC and CBZ–SAC form I showed better photostability than the conventional crystal forms. The order of the photostability of solid forms estimated using FT-IR RAS was almost consistent with the order obtained from the colorimetric measurement (Figure S4, Supplementary Materials). Based on these results, the discoloration on the surface of the tablets was due to the photodecomposition of API. CBZ is transformed into CBZ cyclobutyl dimer and 10,11-epoxide after irradiation by a mercury-vapor lamp [29] and a near-UV fluorescent lamp [22]. These photoproducts are also produced by irradiation with a $D_{65}$ fluorescent lamp, and the surface color of the tablets subsequently changed. The remaining percentages of CBZ form III, CBZ–SUC, and CBZ–SAC form II estimated using FT-IR RAS were approximately 95%, 100%, and 80%, respectively, after 343 h of irradiation. The results obtained using low-frequency Raman spectroscopy showed the same trends as the data obtained using FT-IR RAS. The low-frequency Raman spectroscopy provides complementary information about the photodecomposition of API on the surface of pharmaceutical preparations in a simple and rapid manner.

Molecular mobility correlates with the photodegradation of CBZ. SUC combines with two CBZ molecules through hydrogen bonding [30]. In addition, CBZ–SAC form I forms a homosynthon between two CBZ molecules, hydrogen bonds between SAC molecules and hydrogen bonds between CBZ and SAC [15]. The dimerization of CBZ induced by photoirradiation needs the reorientation of each CBZ molecule, resulting in a satisfying alignment for the reaction. The formation of alternate stacking configurations of CBZ and the corresponding coformers may hinder the approach of CBZ molecules, thus preventing photodimerization, as reported for vitamin K3 by Zhu et al. [17]. CBZ–SAC form II also forms a heterosynthon through hydrogen bonds between CBZ and SAC [15]; however, this polymorphic form exhibited a lower photostability than CBZ itself. According to the thermodynamic study reported by Pagire et al., CBZ–SAC form I is a stable form, CBZ–SAC form II is a metastable form, and these forms are monotropically related polymorphs [31]. In the present study, the $^1$H $T_1$ value of CBZ–SAC form II was significantly lower than the other crystal forms, suggesting that the molecular mobility was sufficiently high to allow for a photoreaction. This higher reactivity and instability of CBZ–SAC form II was supported by the findings from previous reports that CBZ–SAC form II is easily converted to form I in a slurry at room temperature [15,31].

In summary, CBZ–SUC and CBZ–SAC form I are photostable crystal forms, and cocrystallization was confirmed to be an effective approach to improving the photostability of a photolabile drug in the present study.

**Supplementary Materials:** The following are available online at http://www.mdpi.com/2073-4352/9/11/553/s1, Figure S1: Powder X-Ray Diffractometry (PXRD) patterns of CBZ polymorphs, coformers, PMs, and cocrystals; Figure S2: Differential scanning calorimetry (DSC) profiles of CBZ polymorphs, coformers, PMs, and cocrystals; Figure S3: Standard low-frequency Raman spectra of CBZ form III, CBZ–SUC, CBZ–SAC form II, and microcrystalline cellulose; Figure S4: The relationship between the discoloration rate constant and degradation rate constant.

**Author Contributions:** Conceptualization, R.T. and T.F.; formal analysis, R.Y. and R.T.; investigation, R.Y., R.H., and C.T.; writing of original draft preparation, R.Y. and R.H.; writing of review and editing, R.T., C.T., T.K., and T.F.; visualization, R.Y. and R.H.; supervision, S.K., T.S., and T.F.; project administration, R.T. and T.F.

**Funding:** This work was supported in part by JSPS KAKENHI, a Grant-in-Aid for the Scientific Research (C), (grant number: 17K08253) (to T.F.).

**Acknowledgments:** The authors would like to thank Ryota Hamada for providing technical assistance regarding the experiments.

**Conflicts of Interest:** The authors declare no conflicts of interest.

## References

1. Coelho, L.; Almeida, I.F.; Sousa Lobo, J.M.; Sousa e Silva, J.P. Photostabilization strategies of photosensitive drugs. *Int. J. Pharm.* **2018**, *541*, 19–25. [CrossRef]
2. Janga, K.Y.; King, T.; Ji, N.; Sarabu, S.; Shadambikar, G.; Sawant, S.; Xu, P.; Repka, M.A.; Murthy, S.N. Photostability issues in pharmaceutical dosage forms and photostabilization. *AAPS Pharm. Sci. Tech.* **2018**, *19*, 48–59. [CrossRef]
3. Tønnesen, H.H. Formulation and stability testing of photolabile drugs. *Int. J. Pharm.* **2001**, *225*, 1–14. [CrossRef]
4. Grzesiak, A.L.; Lang, M.; Kim, K.; Matzger, A.J. Comparison of the four anhydrous polymorphs of carbamazepine and the crystal structure of form I. *J. Pharm. Sci.* **2003**, *92*, 2260–2271. [CrossRef] [PubMed]
5. Harris, R.K.; Ghi, P.Y.; Puschmann, H.; Apperley, D.C.; Griesser, U.J.; Hammond, R.B.; Ma, C.; Roberts, K.J.; Pearce, G.J.; Yates, J.R.; et al. Structural studies of the polymorphs of carbamazepine, its dihydrate, and two solvates. *Org. Process Res. Dev.* **2005**, *9*, 902–910. [CrossRef]

6. McMahon, L.E.; Timmins, P.; Williams, A.C.; York, P. Characterization of dihydrates prepared from carbamazepine polymorphs. *J. Pharm. Sci.* **1996**, *85*, 1064–1069. [CrossRef]
7. Lowes, M.M.J.; Caira, M.R.; Lotter, A.P.; Van Watt, J.G.D. Physicochemical properties and x-ray structural studies of the trigonal polymorph of carbamazepine. *J. Pharm. Sci.* **1987**, *76*, 744–752. [CrossRef]
8. Kobayashi, Y.; Ito, S.; Itai, S.; Yamamoto, K. Physicochemical properties and bioavailability of carbamazepine polymorphs and dihydrate. *Int. J. Pharm.* **2000**, *193*, 137–146. [CrossRef]
9. Aakeröy, C.B.; Salmon, D.J. Building co-crystals with molecular sense and supramolecular sensibility. *Cryst. Eng. Comm.* **2005**, *7*, 439–448. [CrossRef]
10. Qiao, N.; Li, M.; Schlindwein, W.; Malek, N.; Davies, A.; Trappitt, G. Pharmaceutical cocrystals: An overview. *Int. J. Pharm.* **2011**, *419*, 1–11. [CrossRef]
11. Aakeröy, C.B.; Fasulo, M.E.; Desper, J. Cocrystal or salt: Does it really matter? *Mol. Pharm.* **2007**, *4*, 317–322. [CrossRef]
12. Childs, S.L.; Rodríguez-Hornedo, N.; Reddy, L.S.; Jayasankar, A.; Maheshwari, C.; McCausland, L.; Shipplett, R.; Stahly, B.C. Screening strategies based on solubility and solution composition generate pharmaceutically acceptable cocrystals of carbamazepine. *Cryst. Eng. Comm.* **2008**, *10*, 856–864. [CrossRef]
13. Ullah, M.; Hussain, I.; Sun, C.C. The development of carbamazepine-succinic acid cocrystal tablet formulations with improved in vitro and in vivo performance. *Drug Dev. Ind. Pharm.* **2016**, *42*, 969–976. [CrossRef] [PubMed]
14. Hickey, M.B.; Peterson, M.L.; Scoppettuolo, L.A.; Morrisette, S.L.; Vetter, A.; Guzmán, H.; Remenar, J.F.; Zhang, Z.; Tawa, M.D.; Haley, S.; et al. Performance comparison of a co-crystal of carbamazepine with marketed product. *Eur. J. Pharm. Biopharm.* **2007**, *67*, 112–119. [CrossRef] [PubMed]
15. Porter, W.W., III; Elie, S.C.; Matzger, A.J. Polymorphism in carbamazepine cocrystals. *Cryst. Growth Des.* **2008**, *8*, 14–16. [CrossRef]
16. Wang, J.-R.; Zhou, C.; Yu, X.; Mei, X. Stabilizing vitamin $D_3$ by conformationally selective co-crystallization. *Chem. Commun.* **2014**, *50*, 855–858. [CrossRef]
17. Zhu, B.; Wang, J.-R.; Zhang, Q.; Mei, X. Improving dissolution and photostability of vitamin K3 via cocrystallization with naphthoic acids and sulfamerazine. *Cryst. Growth Des.* **2016**, *16*, 483–492. [CrossRef]
18. Yu, Q.; Yan, Z.; Bao, J.; Wang, J.-R.; Mei, X. Taming photo-induced oxidation degradation of dihydropyridine drugs through cocrystallization. *Chem. Commun.* **2017**, *53*, 12266–12269. [CrossRef]
19. Shinozaki, T.; Ono, M.; Higashi, K.; Moribe, K. A novel drug-drug cocrystal of levofloxacin and metacetamol: Reduced hygroscopicity and improved photostability of levofloxacin. *J. Pharm. Sci.* **2019**, *108*, 2383–2390. [CrossRef]
20. Kaneniwa, N.; Yamaguchi, T.; Watari, N.; Otsuka, M. Hygroscopicity of carbamazepine crystalline powders. *Yakugaku Zasshi* **1984**, *104*, 184–190. [CrossRef]
21. Matsuda, Y.; Teraoka, R.; Sugimoto, I. Comparative evaluation of photostability of solid-state nifedipine under ordinary and intensive light irradiation conditions. *Int. J. Pharm.* **1989**, *54*, 211–221. [CrossRef]
22. Matsuda, Y.; Akazawa, R.; Teraoka, R.; Otsuka, M. Pharmaceutical evaluation of carbamazepine modifications: Comparative study for photostability of carbamazepine polymorphs by using Fourier-transformed reflection-absorption infrared spectroscopy and colorimetric measurement. *J. Pharm. Pharmacol.* **1994**, *46*, 162–167. [CrossRef] [PubMed]
23. Koide, T.; Fukami, T.; Hisada, H.; Inoue, M.; Carriere, J.; Heyler, R.; Katori, N.; Okuda, H.; Goda, Y. Identification of pseudopolymorphism of magnesium stearate by using low-frequency Raman spectroscopy. *Org. Process. Res. Dev.* **2016**, *20*, 1906–1910. [CrossRef]
24. Walker, G.; Römann, P.; Poller, B.; Löbmann, K.; Grohganz, H.; Rooney, J.S.; Huff, G.S.; Smith, G.P.S.; Rades, T.; Gordon, K.C.; et al. Probing pharmaceutical mixtures during milling: The potency of low-frequency Raman spectroscopy in identifying disorder. *Mol. Pharm.* **2017**, *14*, 4675–4684. [CrossRef]
25. Teraoka, R.; Otsuka, M.; Matsuda, Y. Evaluation of photostability of solid-state dimethyl 1,4-dihydro-2,6-dimethyl-4-(2-nitro-phenyl)-3, 5-pyridinedicarboxylate by using Fourier-transformed reflection-absorption infrared spectroscopy. *Int. J. Pharm.* **1999**, *184*, 35–43. [CrossRef]
26. Teraoka, R.; Otsuka, M.; Matsuda, Y. Evaluation of photostability of solid-state nicardipine hydrochloride polymorphs by using Fourier-transformed reflection-absorption infrared spectroscopy—Effect of grinding on the photostability of crystal form. *Int. J. Pharm.* **2004**, *286*, 1–8. [CrossRef]

27. Lubach, J.W.; Xu, D.; Segmuller, B.E.; Munson, E.J. Investigation of the effects of pharmaceutical processing upon solid-state NMR relaxation times and implications to solid-state formulation stability. *J. Pharm. Sci.* **2007**, *96*, 777–787. [CrossRef]
28. Hudson, S.P.; Owens, E.; Hughes, H.; McLoughlin, P. Enhancement and restriction of chain motion in polymer networks. *Int. J. Pharm.* **2012**, *430*, 34–41. [CrossRef]
29. Robson, J.K.; Sharples, D. Photoirradiation products of cyproheptadine and carbamazepine. *J. Pharm. Pharmacol.* **1984**, *36*, 843–844. [CrossRef]
30. Shenmin, X.; Dang, L.; Wei, H. Solid-liquid phase equilibrium and phase diagram for the ternary carbamazepine-succinic acid-ethanol or acetone system at (298.15 and 308.15) K. *J. Chem. Eng. Data* **2011**, *56*, 2746–2750. [CrossRef]
31. Pagire, S.K.; Jadav, N.; Vangala, V.R.; Whiteside, B.; Paradkar, A. Thermodynamic investigation of carbamazepine-saccharin co-crystal polymorphs. *J. Pharm. Sci.* **2017**, *106*, 2009–2014. [CrossRef] [PubMed]

© 2019 by the authors. Licensee MDPI, Basel, Switzerland. This article is an open access article distributed under the terms and conditions of the Creative Commons Attribution (CC BY) license (http://creativecommons.org/licenses/by/4.0/).

*Article*

# Structural and Reactivity Analyses of Nitrofurantoin–4-dimethylaminopyridine Salt Using Spectroscopic and Density Functional Theory Calculations

Eram Khan [1], Anuradha Shukla [1], Karnica Srivastava [1], Debraj Gangopadhyay [1], Khaled H. Assi [2], Poonam Tandon [1,*] and Venu R. Vangala [2,*]

1 Department of Physics, University of Lucknow, Lucknow 226 007, India
2 Centre for Pharmaceutical Engineering Science and School of Pharmacy and Medical Sciences, University of Bradford, Bradford BD7 1DP, UK
* Correspondence: poonam_tandon@yahoo.co.uk (P.T.); V.G.R.Vangala@bradford.ac.uk (V.R.V.); Tel.: +91-522-270840 (P.T.); +44 127423 6116 (V.R.V.)

Received: 4 July 2019; Accepted: 6 August 2019; Published: 9 August 2019

**Abstract:** Pharmaceutical salt, nitrofurantoin–4-dimethylaminopyridine (NF-DMAP), along with its native components NF and DMAP are scrutinized by FT-IR and FT-Raman spectroscopy along with density functional theory so that an insight into the H-bond patterns in the respective crystalline lattices can be gained. Two different functionals, B3LYP and wB97X-D, have been used to compare the theoretical results. The FT-IR spectra obtained for NF-DMAP and NF clearly validate the presence of C33–H34$\cdots$O4 and N23–H24$\cdots$N9 hydrogen bonds by shifting in the stretching vibration of –NH and –CH group of DMAP$^+$ towards the lower wavenumber side. To explore the significance of hydrogen bonding, quantum theory of atoms in molecules (QTAIM) has been employed, and the findings suggest that the N23–H24$\cdots$N9 bond is a strong intermolecular hydrogen bond. The decrement in the HOMO-LUMO gap, which is calculated from NF $\to$ NF-DMAP, reveals that the active pharmaceutical ingredient is chemically less reactive compared to the salt. The electrophilicity index ($\omega$) profiles for NF and DMAP confirms that NF is acting as electron acceptor while DMAP acts as electron donor. The reactive sites of the salt are plotted by molecular electrostatic potential (MEP) surface and calculated using local reactivity descriptors.

**Keywords:** Nitrofurantoin–4-dimethylaminopyridine (NF-DMAP) salt; DFT study; HOMO-LUMO; reactivity descriptors; hydrogen bonding

---

## 1. Introduction

The active pharmaceutical ingredient (API) is a drug or a chemical liable for the pharmacological and therapeutic activities in the body. It is the key ingredient that treats a disease or disorder [1]. In recent years, pharmaceutical salts have drawn considerable attention owing to their impending potential uses in pharmacy and biomedical science [2]. The market value of a drug can be drastically reduced due to its poor physicochemical properties, which in turn can cause the demand to replace it. This is the major reason why there is an increased interest in the physicochemical properties upgradation methods such as salt formation [3]. Pharmaceutical salts are ionizable drugs that have been combined with a counter-ion to form a neutral compound. Salts are stable, and due to the presence of ionic bond, they are highly soluble in the polar solvents that include water [4]. The presence of ionizable groups in the molecule is an essential requirement for the formation of salts. An API can be either in the form of cation (approximately 75% of pharmaceutical salts) or anion, and the counter molecule is called as coformer, which can be either in the form of cation or anion. By changing the coformer of a pharmaceutical salt, the physicochemical properties of the drug can be modulated [5–9]. Converting a neutral API into a salt may enhance its chemical stability, solubility and bioavailability, and the solid form of the drug is made easier to administer. In medicinal treatment, nearly half of the drugs used today are given as salt forms. This implies that the saltification of a drug molecule is a substantial and beneficial stage in drug development.

Nitrofurantoin (NF), a drug of nitrofurans group, is an antibiotic drug to cure urinary tract (kidney and bladder) infections (UTI) and also provides chronic cure against recurrent infections. As antibacterial resistance to NF is rare, it is used for long term treatment of UTIs [10,11]. NF prevents several bacterial enzyme systems (gram-positive and gram-negative bacteria) and has broad antibacterial activity [12]. The dissolution rate of NF in water is low [10], and also its bioavailability drops upon storage [13–16]. Due to these paucities and to enhance the physicochemical stability (solubility and bioavailability) of NF tablets, preparation of pharmaceutical salt of NF could potentially present significant opportunities [17].

In the present work, pharmaceutical salt nitrofurantoin–4-dimethylaminopyridine (NF-DMAP) is studied where NF is an API in anionic form and DMAP a coformer, which is one of the derivatives of pyridine in cationic form. DMAP is more basic compared to pyridine with the electron releasing substituents (methyl groups) on the amino-nitrogen atom, which is disposed in para-position to that of pyridyl group. The crystal structure of NF-DMAP is reported by one of the authors of this manuscript [18]. NF-DMAP salt has conceived the monoclinic crystal system with $P2_1/c$ space group [18].

Salt formation is a promising avenue that is used to increase the therapeutic efficacy of APIs. In order to examine the changes/alteration from API (NF) to salt (NF-DMAP), we have computationally calculated the chemical reactivity, vibrational properties, effect of hydrogen bonding in NF-DMAP.

In this work, a complete vibrational study of NF-DMAP salt and NF by Raman and infrared (IR) spectral analysis combined with density functional theory (DFT) approach has been performed. It should be mentioned that vibrational spectroscopy is the most accepted analytical tool, and DFT is the utmost acknowledged theoretical approach to study the molecular structure of different molecular systems [19–24]. In order to attain a comprehensive analysis of the hydrogen bond patterns, both DFT and spectroscopic methods have been used. From the computational studies, the results obtained using two functionals (B3LYP and wB97X-D) employing a single basis set 6-311++G(d,p) are compared to understand the geometry and reactivity of NF-DMAP salt. Natural bond orbital (NBO) analysis is used to examine the strength of intermolecular H-bonds, and QTAIM approach confirms the results. Moreover, the performance of two descriptors (global and local reactivity descriptors) has been examined to explain the subtle changes in the properties of nitrofurantoin–4-dimethylaminopyridine from NF and DMAP.

## 2. Experimental Details

NF (β-form) and DMAP were obtained from Sigma-Aldrich and acetonitrile is of analytical grade and used as received. As per literature, solution crystallisation of NF and DMAP was performed using the equimolar amounts of NF and DMAP using acetonitrile to obtain crystals of NF-DMAP salt [18].

Shimadzu IR Affinity-1S FTIR spectrophotometer has been employed to record IR spectra of NF-DMAP salt (at 34 °C).

The transmission infrared spectrum of NF was acquired by using an FT-IR spectrometer (Bio-Rad, FTS 3000 MX IR spectrometer, Singapore) (at 30 °C) [25].

The FT-Raman spectrum of NF-DMAP salt was documented on a Bruker IFS 55 EQUINOX (at 30 °C) [25].

The dispersive-Raman microscope employed in the study of NF was a JY Horiba LabRAM HR equipped with a confocal microscope (at 29 °C) [25].

## 3. Computational Details

Initially, energy and optimized electronic structure of NF-DMAP salt and its precursor components were computed by the DFT method [26] using the Gaussian 09 suite [27]. 6-311++G(d,p) [28,29] basis set with B3LYP [30–32] and wB97X-D functional [33] (uses a variation of Grimme's D2 dispersion model) was used for the calculations. An inclusive set of 120 internal coordinates were well-defined using Pulay's recommendations for normal mode analysis [34]. The vibrational assignments of the normal modes were recommended based on the potential energy distribution (PED) calculated using the program Gar2Ped [35]. Pictorial representation of molecular geometries and substantiation of calculated data were prepared with the program GaussView [36]. Topological parameters at bond critical points (BCP) were analysed within the framework of the QTAIM with the use of AIM2000 program package [37]. The details of the theoretical background are given in the Supplementary Material.

## 4. Results and Discussion

### 4.1. Geometry Optimization and Energies

The crystal structure determination of NF-DMAP salt suggests its crystallization in the monoclinic space group $P2_1/c$ with one molecule each of NF$^-$ and DMAP$^+$ in the asymmetric unit [18]. The reported lattice parameters for the NF-DMAP are $a$ = 8.616(4) Å, $b$ = 26.655(11) Å and $c$ = 7.029(3) Å; $\alpha$ = 90°, $\beta$ = 100.722(9) and $\gamma$ = 90 [18]. Transfer of a proton took place from the imide functional group (N–H) of NF to the pyridyl–N of DMAP. In the crystallographic structure of the salt, four molecules of NF-DMAP (monomer) are present in a unit cell (Figure S1), but NF$^-$ is forming H-bond with one DMAP$^+$ only (Figures S2 and S3). No H-bond interactions are present between one unit of NF-DMAP (monomer) and other (Figure S3). Different units of NF-DMAP (monomer) within the unit cell are arranged in layer form and interact with each other via weak interactions, i.e., π-π stacking (non-classical interaction) (Figure S4). This is the reason why, here, calculations have been performed on the NF-DMAP (1:1, monomer) salt which has taken into account all the H-bond interactions. The three crystal systems considered in this study, NF-DMAP (1:1), NF (monomer) and DMAP (monomer), are known [15,18,38]. Hence, their crystallographic data have been served to obtain the initially optimized structure with energy minima. These three structures are optimized at the B3LYP and wB97X-D theory of DFT with 6-311++G(d,p) as basis set.

In the present study, we have probed the performance of density functionals, one of which is the "standard" functional that does not include dispersion (B3LYP) whereas the other one takes into account dispersion (wB97X-D) in reproducing molecular structures. The primary difference between B3LYP and wB97X-D is the addition of a semi empirical dispersion term which causes the weaker London forces. Different theories are used for different calculation schemes (B3LYP and wB97X-D), which is why they have led to slightly different 3D molecular structures. The main purpose of our work is to compare two different functional in DFT to understand the geometry and reactivity of NF-DMAP salt. The ground state optimized geometries of NF-DMAP (B3LYP), NF-DMAP (wB97X-D), NF (wB97X-D and B3LYP) and DMAP (wB97X-D and B3LYP) are shown in Figures 1 and 2, Figures S5 and S6 respectively. The ground state energy of the NF-DMAP (1:1) salt is computed by using B3LYP and wB97X-D; methods are −1286.054068 and −1285.61456 Hartrees, respectively. The value of this energy for NF and DMAP calculated using wB97X-D (B3LYP) are −903.40623 (−903.71326) and −382.23392 (−382.36016) Hartrees, respectively. The molecular geometries obtained using B3LYP functional are more stable.

Optimized geometrical parameters of NF-DMAP and NF calculated by B3LYP and wB97X-D methods are presented in Table S1, combined with the crystallographic data of both salt and API [15,18]. The molecular geometries of NF-DMAP calculated using B3LYP and wB97X-D methods are compared with the geometry of a single molecule of NF. When the geometrical parameters of NF-DMAP calculated using both the functionals and the experimental values are compared, it shows that the calculations are comparable. Minor changes, however, are seen in the geometrical parameters involving in hydrogen bonding. Intermolecular H-bonds (N23–H24···N9 and C33–H34···O4) are present within the salt. The calculated length of C33–H34 bond of DMAP and NF-DMAP using wB97X-D (B3LYP) are 1.08735 (1.08713) Å and 1.09207 (1.08944) Å, respectively. In case of NF-DMAP, this C33–H34 bond has slightly greater length as compared to DMAP only, as in DMAP, this bond is free while it is involved in H-bond in case of NF-DMAP. When comparison is done between NF-DMAP (1:1) and NF (monomer) for the C18–O4 bond, a difference of 0.0252 Å (B3LYP) and 0.0251 Å (wB97X-D) has been observed. In case of salt, C33–H34···O4 H-bond is formed while in NF (API only) this C18–O4 bond is free. A difference of 0.0283 Å has been observed between experimental bond length (N9-C18) of salt and API, as in NF, hydrogen atom is attached to N9 atom whereas this hydrogen is transferred to pyridyl-N23 of coformer resulting in N23–H24 bond. As a result of transferring of H atom from N9 to N23, shortening in the bond length N9-C18 occurs in case of salt. The bond length calculated using B3LYP is quite similar to the observed value, which means for the present study B3LYP is giving plausible results for calculating parameters involved in hydrogen bonding [24]. The angle values are also almost the same when calculated by two functionals; however, small differences are observed between values involved in H-bonding. Small deviations are found in dihedral angles involved in intermolecular H-bonding (see Figures 1 and 2).

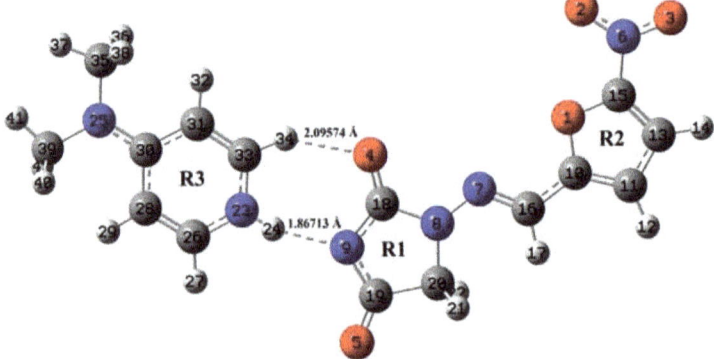

**Figure 1.** Optimized structure of NF-DMAP using B3LYP method.

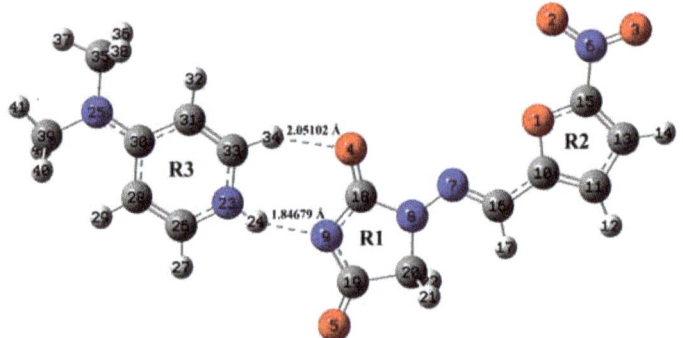

**Figure 2.** Optimized structure of NF-DMAP using wB97X-D method.

### 4.2. Vibrational Assignment

The total number of atoms in API (NF) and salt (NF-DMAP) are 23 and 42, which means that these can give 63 and 120 (3N−6) normal modes, respectively. The theoretical and experimental vibrational of NF-DMAP and NF calculated at B3LYP and wB97X-D methods and their assignments using PED are given in the Tables S2 and S3 (for NF-DMAP) and the Tables S4 and S5 (for NF), respectively.

The relative study of experimental and calculated (scaled) IR and Raman spectra of NF is given in the Figures S7 and S8.

The calculated Raman and IR intensities were used to convolute each predicted vibrational mode with a Lorentzian line shape (FWHM = 4 cm$^{-1}$) to produce simulated spectra. Comparison between experimental and theoretical IR spectra for NF-DMAP (calculated using wB97X-D and B3LYP) in the region 3500–400 cm$^{-1}$ is presented in Figure 3. Calculated (scaled) and experimental Raman spectra of salt are shown in Figure 4. As the comparison is made between the calculated (gaseous phase) and experimental (solid state) spectra and the anharmonicity effects are also not included, the calculated wave numbers found in the present study are somewhat greater than the observed ones. Consequently, the computed wavenumbers are reduced by 0.980 and 0.991 for wB97X-D and B3LYP functionals, respectively, to eliminate anharmonicity existing in real system [39–41]. The main motive to discuss the vibrational modes corresponding to the bands affected by hydrogen bonds in the salt is to give information related to the functional groups, i.e., whether they are bonded or non-bonded, and to confirm the transfer of proton from NF$^-$ to DMAP$^+$ and also to know the sites which took part in forming hydrogen bond between NF (API) and DMAP (coformer), which are responsible for making NF-DMAP salt.

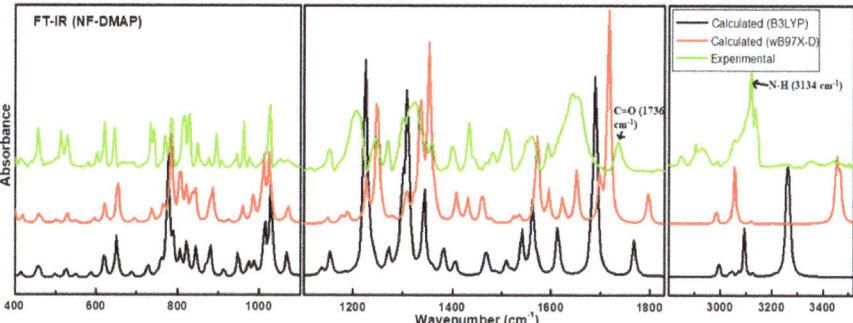

**Figure 3.** Experimental and theoretical IR spectra for NF-DMAP in the range 3600–400 cm$^{-1}$.

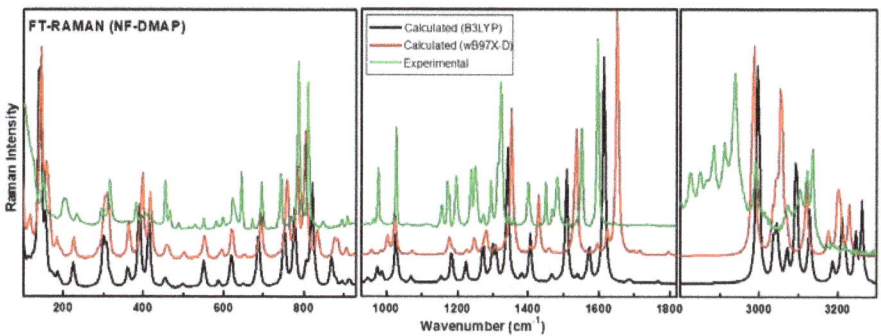

**Figure 4.** Experimental and theoretical Raman spectra for NF-DMAP in the range 3600–100 cm$^{-1}$.

### 4.3. Vibrational Wavenumbers Involved in Hydrogen Bonding

The N–H group of pyridyl ring is H-bonded with N9 atom of the hydantoin ring. Because of the presence of N23–H24···N9 H-bond, change in bondlengths and wavenumbers associated with it occurred. The position of N23–H24 stretching band depends on the potency of the formation of H-bond. In the observed FT-IR spectrum of NF-DMAP salt, the N23–H24 stretch is seen at 3134 cm$^{-1}$, whereas it is calculated as 3458 and 3263 cm$^{-1}$, using wB97X-D and B3LYP level of theory, respectively. The peak of free N-H bond falls in the range 3600–3400 cm$^{-1}$ [39–41]. However, the N–H peak in

NF-DMAP is somewhat lower than this range. In neutral DMAP, which is not in cationic form, no N–H stretching peak occurs in the vibrational spectra. The lowering of N–H stretching vibrations in the observed spectra can be endorsed to the intermolecular N–H⋯N interaction [42,43].

The stretching vibrations of the carbonyl (C=O) group usually cause the bands in the region 1600–1800 cm$^{-1}$ in aromatic compounds [44–46]. The stretching vibration of the C=O4 group is observed at 1736 cm$^{-1}$ in IR spectrum and calculated at 1797/1768 cm$^{-1}$ using wB97X-D/B3LYP. The same mode is calculated at 1639/1547 cm$^{-1}$ using wB97X-D/B3LYP in the case of NF and found at 1566/1563 cm$^{-1}$ in the IR/Raman spectra.

Moreover, amongst different C-H bonds of ring 3, one (C-H34) is H bonded while others are free resulting in C-H34 bond elongation. Because of this, C-H34 stretching wavenumber is calculated at 3055/3094 cm$^{-1}$ (wB97X-D/B3LYP) while free C-H stretching wavenumber occurs at higher wavenumber [34]. This proves that the C-H34 group of DMAP$^+$ plays a part in intermolecular H-bonding with C=O4 of NF$^-$. For some molecular systems, wB97X-D gives good results; however, for other systems, B3LYP affords better results [23,24]. It may be concluded that wavenumbers calculated using B3LYP method match well with the observed values showing B3LYP is better functional than wB97X-D for the studied molecular system.

### 4.4. Quantum Theory of Atoms in Molecules (QTAIM) Calculations for Hydrogen Bonding

QTAIM is a technique for estimation and comparison of properties (especially chemical bonds) of atoms and molecules [37]. This theory creates the spatial partition of atoms. The presence of critical points describes the presence of a bond amongst two nuclei in QTAIM. With the purpose of having a perception into a region of a system, this method has been used. Geometrical and topological parameters are appropriate means to exemplify the nature and strength of H-bonds. The geometrical conditions for the occurrence of hydrogen bonds are given by Koch and Popelier centered on QTAIM [47]. The presence of H-bonds are also supported by Rozas et al. [48], which classifies these interactions as follows: (i) strong H-bonds have Laplacian of electron density ($\nabla^2 \rho_{BCP}$) < 0, total electron energy density $H_{BCP}$ < 0 and have covalent character; (ii) medium H-bonds have ($\nabla^2 \rho_{BCP}$) > 0, $H_{BCP}$ < 0 and have partially covalent character and (iii) weak H-bonds have ($\nabla^2 \rho_{BCP}$) > 0 and $H_{BCP}$ > 0 and have electrostatic character.

The molecular graphs of NF-DMAP using the AIM program at the wB97X-D/6-311++G(d,p) and B3LYP/6-311++G(d,p) level are shown in Figures S9 and S10, respectively. Topological, geometrical and energy parameters for intermolecular hydrogen bonds of interacting atoms at the wB97X-D/6-311++G(d,p) and B3LYP/6-311++G(d,p) level are listed in Table 1. The geometrical parameters for hydrogen bonds in NF-DMAP are given in Table S6, and bond contributions to atomic net charges are tabulated in Table S7. According to Rozas theory, all the H-bonds are medium in nature and have partial covalent character as they have ($\nabla^2 \rho_{BCP}$) > 0, $H_{BCP}$ < 0. In this contribution, hydrogen bond energy ($E_{HB}$) is also calculated using the QTAIM theory. $E_{HB}$ is correlated to $V_{BCP}$ by the relation; $E_{HB} = 1/2(V_{BCP})$. Amongst all the H-bonds N23–H24⋯N9 is the strongest H-bond present between NF$^-$ and DMAP$^+$.

**Table 1.** Geometrical (calculated bond length) and topological parameters for hydrogen bonds of interacting atoms of NF-DMAP: Electron density ($\rho_{BCP}$), Laplacian of electron density ($\nabla^2 \rho_{BCP}$), electron kinetic energy density ($G_{BCP}$), electron potential energy density ($V_{BCP}$), total electron energy density ($H_{BCP}$) at bond critical point (BCP) and hydrogen bond energy ($E_{HB}$).

| Interactions | Bond Length (Å) | $\rho_{BCP}$ (a.u.) | $\nabla^2 \rho_{BCP}$ (a.u.) | $G_{BCP}$ (a.u.) | $V_{BCP}$ (a.u.) | $H_{BCP}$ (a.u.) | $E_{HB}$ (kcal mol$^{-1}$) |
|---|---|---|---|---|---|---|---|
| | | | B3LYP | | | | |
| N23–H24⋯N9 | 1.86713 | 0.03701 | 0.11551 | 0.00107 | −0.03103 | −0.02996 | −9.7358 |
| C33–H34⋯O4 | 2.09574 | 0.02015 | 0.06811 | −0.00216 | −0.01270 | −0.01486 | −3.9847 |
| | | | wB97X-D | | | | |
| N23–H24⋯N9 | 1.84679 | 0.03882 | 0.12267 | 0.00156 | −0.03378 | −0.03222 | −10.5986 |
| C33–H34⋯O4 | 2.05102 | 0.02196 | 0.07461 | −0.00216 | −0.01432 | −0.01648 | −4.4926 |

## 4.5. Natural Bond Orbital (NBO) Analysis

NBO analysis deals with the tactic that studies inter and intramolecular orbital interaction in a molecule, predominantly charge transfer or conjugative interactions in the molecular system. This is executed in view of all likely interactions amongst donor and acceptor NBOs and valuing their energetic significance by second-order perturbation theory. A stabling donor-acceptor interaction is obtained when the delocalization of electron density amongst occupied Lewis type as well as unoccupied non-Lewis NBO orbitals occurs. The strength of the interaction and the amount of conjugation of the system can be predicted by the $E^{(2)}$ value.

NBO analysis is accomplished on NF-DMAP using wB97X-D/6-311++G(d,p) and B3LYP/6-311++G(d,p) methods with the purpose of interpreting the intermolecular rehybridization and delocalization of electron density within the NF-DMAP salt. The Tables S8 and S9 present the particular electron donor orbital, acceptor orbital and the interacting stabilization energy for the NF-DMAP following from the second-order perturbation theory.

In NF-DMAP, the significant interaction within unit 1 from n(3) O3→π*O2-N6 has stabilization energy of 148.50 kcal/mol, n(2) N9→ π*O4-C18, and π*O5-C19 has 84.89 kcal/mol and 76.95 kcal/mol, respectively, and consequently, they provide sturdier stabilization to the structure (Table S7). From Table S7, it is prominent that the maximum occupancies 1.98661 and 1.98618 are found for πO4-C18 and πO5–C19, respectively. Therefore, the results suggest that the πO4-C18 and πO5–C19 are handled by π-character of the hybrid orbitals.

The strongest interactions, computed in similar interaction energy, are the electron donations from a lone pair orbital on the oxygen atom, electron donating from n(3)O3 to π*O2–N6, tends to high stabilization energy of 227.86 kcal/mol using wB97X-D method (Table S8). The strong intra-molecular hyperconjugation interaction of the πC26–C28/πC31–C33 to the n(1) N23 bond in the ring leads to stabilization of ring present in the coformer.

In NF-DMAP, charge transfer from NF⁻ unit (1) to DMAP⁺ unit (2) due to n(1) N9→ σ*N23–H24 stabilized the molecule up to 18.66 and 22.72 kcal/mol, calculated using B3LYP and wB97X-D methods, respectively, and this proves the existence of classical interaction N23–H24···N9.

## 4.6. Chemical Reactivity

### 4.6.1. Frontier Molecular Orbitals (FMOs)

Frontier molecular orbitals (FMOs) and their properties are used for describing various kinds of reactions and for calculating the most reactive site in a conjugated system. To reveal the chemical reactivity and biological activity of a complex, the HOMO and LUMO energy and their gap are calculated. HOMO, which may be believed as the outer orbital holding electrons, has a tendency to donate the electrons as e⁻ donor and henceforth the ionization potential is directly associated with the energy of the HOMO. In contrast, LUMO may take electrons and the LUMO energy is directly associated with electron affinity [49]. In NF-DMAP, the HOMO is located over the complete NF. Then again, in the API alone, the LUMO is located over the complete molecule except for hydantoin ring.

A molecule having a lesser HOMO-LUMO gap is more polarizable and is usually related to low kinetic stability and high chemical reactivity [49,50]. Thus, the softness corresponds to the HOMO-LUMO gap. The smaller the HOMO-LUMO energy gap the softer the molecule or a multicomponent crystal. The HOMO-LUMO energy gaps of API and salt calculated at wB97X-D (B3LYP) are shown in Figures 5 and 6. On-going from API to salt, the energy gap reduces for both the functionals, which infers that the chemical reactivity of salt is more than API. The energy gap of NF-DMAP salt is less than the NF (API) and previously studied cocrystal of NF, nitrofurantoin-melamine monohydrate (NF-MELA-H₂O) [24]. Therefore, salt is showing better chemical reactivity than cocrystal and API.

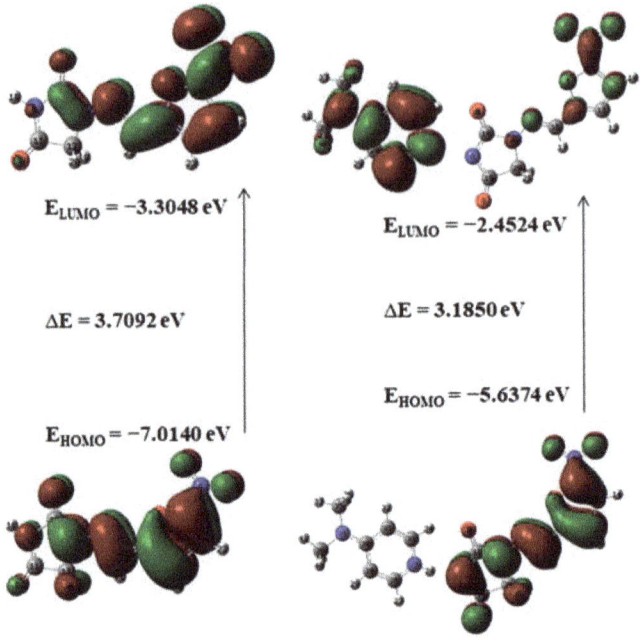

**Figure 5.** The HOMO-LUMO energy gap of NF and NF-DMAP calculated at the B3LYP level of theory.

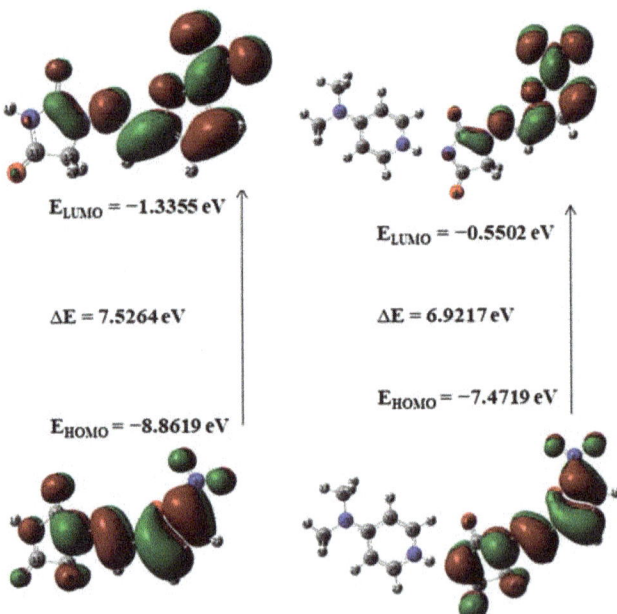

**Figure 6.** The HOMO-LUMO energy gap of NF and NF-DMAP calculated at the wB97X-D level of theory.

### 4.6.2. Molecular Electrostatic Potential (MEP) Analysis

MEP surface proposes a visual technique to comprehend the comparative polarity of the compound [51,52]. It provides a pictorial method to recognize the relative polarity of the compound. MEP is extensively employed as a reactivity plot presenting most plausible areas for the electrophilic attack. The values and spatial dispersal of MEP are in reality accountable for the chemical performance of an agent in a reaction. The MEP maps permit to visualize differently charged regions of a complex. Charge distributions determine how two molecules interact with each another. The Figures S11 and S12 depict the MEP of NF and DMAP mapped using B3LYP method, which explains the charge distributions of the molecule in 3-D. The figure exhibits different colors on the surface that represent different values of the electrostatic potential. Green, red, and blue colors represent regions of zero, most electronegative and most positive electrostatic potential, respectively. Potential decreases in the order blue > green > yellow > orange > red. Red color specifies the high repulsion and blue specifies the high attraction.

The imide group of hydantoin ring of NF is a major nucleophilic centre (blue in colour) while the pyridyl-N of DMAP is a major electrophilic centre (red in colour). Due to the proton transfer from the imide group of NF to the pyridyl-N of DMAP, the reduction in electrostatic potential around these sites (green in colour) takes place, resulting in the formation of NF-DMAP salt.

Here pictorial representation of the atomic regions of NF-DMAP, which are prone to electrophilic or nucleophilic attack, has been done. Regions of negative potential are generally related with the lone pair of electronegative atoms. MEP map of NF-DMAP along with electrostatic potential V (r) and point charges (e), presented in Figure S13 (B3LYP), shows that the negative region is mainly localized over the C=O groups of hydantoin ring and $NO_2$ group of NF anion as these sites are related with the lone pair of oxygen atoms. The maximum positive region is localized on the CH group of ring 3 and $CH_3$ group of $DMAP^+$. Therefore, additional bonding will prevail on these sites.

### 4.6.3. Global Reactivity Descriptors

For comprehending several aspects of pharmacological sciences comprising drug design and the potential eco-toxicological physiognomies of the drugs, numerous new chemical reactivity descriptors have been suggested.

The conceptual DFT commendably dealt with the understanding of chemical reactivity and site selectivity of the molecular systems. Global reactivity descriptors; chemical potential ($\mu$), electronegativity ($\chi$), global hardness ($\eta$), global softness (S) and global electrophilicity index ($\omega$) are extremely effective in calculating global chemical reactivity trends. These descriptors refer to the overall stability of the molecule. Contrariwise, the local equivalent refers to the site reactivity and selectivity [49,50]. The HOMO-LUMO energy gap, $\chi$, $\mu$, $\eta$, S and $\omega$ for API, coformer, salt using wB97X-D and B3LYP methods are listed in Table 2. It is seen that the $\mu$ of NF-DMAP is negative, and it means that it does not decompose suddenly into the components it is made up of. The $\eta$ specifies the conflict towards the distortion of electron cloud of chemical systems under small disturbance that comes across during the chemical process. According to the definition of $\omega$, it measures the vulnerability of chemical species to take electrons. Low values of $\omega$ recommend a good nucleophile; however, greater values specify the existence of good electrophiles [53]. At the B3LYP and wB97X-D level, the electrophilicity index values are following the order API < salt. The last result suggests that the NF-DMAP is the chemically most reactive species. Furthermore, it can be said that the values of these descriptors depend on the level of theory used.

The value of electrophilic charge transfer (ECT) using two functionals B3LYP and wB97X-D comes out to be 1.5879 and 0.1053 for the reactant molecules NF and DMAP which directs that charge travels from NF (API) to DMAP (coformer) during the formation of salt as the value of ECT is greater than zero (Table 2). The smaller value of $\mu$ and larger value of $\omega$ for NF furthermore approve its electrophilic actions. Similarly, larger value of $\mu$ and a smaller value of $\omega$ for DMAP approve its nucleophilic actions. Therefore, NF acts as an $e^-$ acceptor and DMAP as an $e^-$ donor.

**Table 2.** Calculated $E_{HOMO}$, $E_{LUMO}$, energy gap ($E_L$–$E_H$), chemical potential ($\mu$), electronegativity ($\chi$), global hardness ($\eta$), global softness (S) and global electrophilicity index ($\omega$) at 298.15 K for NF, DMAP and NF-DMAP using wB97X-D and B3LYP level of theory.

| Molecule | $E_H$ (eV) | $E_L$ (eV) | $E_L$–$E_H$ (eV) | $\chi$ (eV) | $\mu$ (eV) | $\eta$ (eV) | S (eV$^{-1}$) | $\omega$ (eV) | $\Delta N_{max}$ | ECT |
|---|---|---|---|---|---|---|---|---|---|---|
| | | | | B3LYP | | | | | | |
| NF | −7.0140 | −3.3048 | 3.7092 | 5.1590 | −5.1594 | 1.8546 | 0.2696 | 7.1766 | 2.7819 | |
| DMAP | −5.9664 | −0.5276 | 5.4388 | 3.2470 | −3.2470 | 2.7194 | 0.1839 | 1.9385 | 1.1940 | 1.5879 |
| NF-DMAP | −5.6374 | −2.4524 | 3.1850 | 4.0449 | −4.0449 | 1.5925 | 0.3139 | 5.1369 | 2.5400 | |
| | | | | wB97X-D | | | | | | |
| NF | −8.8619 | −1.3355 | 7.5264 | 5.0990 | −5.0987 | 3.7632 | 0.1329 | 3.4541 | 1.3549 | |
| DMAP | −7.8984 | −0.8765 | 7.0219 | 4.3870 | −4.3875 | 3.5110 | 0.1424 | 2.7414 | 1.2496 | 0.1053 |
| NF-DMAP | −7.4719 | −0.5502 | 6.9217 | 4.0110 | −4.0110 | 3.4608 | 0.1445 | 2.3243 | 1.1590 | |

### 4.6.4. Local Reactivity Descriptors

With the purpose of determining the accurate distribution of the active sites of NF-DMAP; Fukui function (FF, $f_k^+$, $f_k^-$ and $f_k^0$) values are calculated. It is significant to highlight that the FF values are dependent on the type of charges used. Here, Hirshfeld charges are taken for the calculations, which were calculated using B3LYP/6-311++G(d,p) and wB97X-D/6-311++G(d,p) methods. The Tables S10 and S11 show the Fukui functions, local softnesses and local electrophilicity indices for certain atomic sites present in NF-DMAP and NF. Here $f_k^+/f_k^-$ and $f_k^-/f_k^+$ ratios are also calculated in addition to FF ($f_k^+$, $f_k^-$), local softness ($S_k^+$, $S_k^-$) and electrophilicity indices ($\omega_k^+$, $\omega_k^-$) as these ratios reveal different reactive atomic sites having high values for both $f_k^+$ and $f_k^-$. In case of wB97X-D level of theory, it has been established that the carbon atom C16 has a higher $f_k^+$ value which means this atomic site is bound to nucleophilic attack. Whereas C15 has the highest value of $f_k^-$ value which specifies that this atomic site is prone to electrophilic attack. While in case of B3LYP level of theory, nitrogen atom N8 is suitable for nucleophilic attack and oxygen atom O3 is suitable for electrophilic attack.

## 5. Conclusions

The present work focused on studying the hydrogen bonding interactions, structure–reactivity relationships between a pharmaceutical salt and coformers that include active ingredient for the development of the new drugs. In this contribution, a salt of an antibiotic drug, nitrofurantoin-4-dimethylaminopyridine (NF-DMAP), along with its precursor components (NF and DMAP), was characterized by vibrational spectroscopy (FT-IR and Raman) and DFT using two functionals B3LYP and wB97X-D. To exploit the potential of hydrogen bonding and reactivity, QTAIM, and to understand electrostatic potential Frontier, molecular analyses have been performed. The findings suggest that the geometric parameters calculated at B3LYP level represent a worthy approximation to the crystallographic values as compared to wB97X-D level of computations. On comparing the calculated bond lengths of C33–H34 bond, which is involved in H-bonding, and C26–H27 bond, which is free, a difference of 0.00714 (0.00978) Å has been observed using B3LYP(wB97X-D) functional. This difference is because of the interaction of C33–H34 bond of DMAP$^+$ with carbonyl group (C=O4) of NF$^-$ resulting in the formation C33–H34···O4–C18 hydrogen bond and in the elongation of C33–H34 bond in NF-DMAP. Formation of C33–H34··· O4–C18 H-bond is also confirmed by the spectroscopic studies. The N–H stretching vibration in NF-DMAP is observed at 3134 cm$^{-1}$, which confirms the formation of N23–H24···N9 H-bond. The potency of this H-bond is also calculated by NBO and QTAIM and comes out to be −9.6009 kcal mol$^{-1}$ (using B3LYP theory). Frontal molecular orbital analysis gives an indication that charge transfer takes place within NF-DMAP salt. Reduction in the electrostatic potential of MEP surface of NF-DMAP also confirms the proton transfer from NF to DMAP. The HOMO-LUMO gap decreases on moving from NF to NF-DMAP, which infers NF is chemically less reactive than NF-DMAP. Moreover, salt is showing better chemical reactivity than previously published cocrystal (NF-MELA-H$_2$O). The ECT value is greater than zero indicating that charge is transferred from NF to DMAP. From the Fukui function, the most reactive site for electrophilic attacks is observed on the oxygen (O3) of nitro group of NF$^-$. The more reactive site for the nucleophilic attack is located on the nitrogen N8 using B3LYP functional, which is also visible from the MEP surface of NF-DMAP. Thus, the present work significantly enhances the knowledge of structural and chemical reactivity of NF-DMAP, which possibly offers analogies for studying the structure–reactivity relationships and for the development of the new drugs.

**Supplementary Materials:** The following are available online at http://www.mdpi.com/2073-4352/9/8/413/s1. Figure S1: H-bond interactions within the unit cell of NF-DMAP; Figure S2: Intermolecular H-bond present within NF-DMAP.; Figure S3: All the intermolecular H-bonds present within the unit cell; Figure S4: Short contacts present between two molecules (monomer) of NF-DMAP; Figure S5: Optimized structure of NF using wB97X-D and B3LYP level of theories; Figure S6: Optimized structure of DMAP using wB97X-D and B3LYP level of theories; Figure S7: Comparison of experimental and calculated (scaled) IR spectra of NF; Figure S8: Comparison of experimental and calculated (scaled) Raman spectra of NF; Figure S9: Molecular graph of the NF-DMAP using B3LYP level of theory; Figure S10: Molecular graph of the NF-DMAP using wB97X-D level of theory; Figure S11: Molecular electrostatic potential (MEP) for NF using B3LYP level of theory; Figure S12: Molecular electrostatic potential (MEP) for DMAP using B3LYP level of theory; Figure S13: Molecular electrostatic potential (MEP) along with electrostatic potential V (r) and point charges (e) for NF-DMAP using B3LYP level of theory. Table S1: The experimental geometric parameters of NF-DMAP and NF and calculated geometric parameters of NF-DMAP and NF using B3LYP/6-311++g(d,p) and wB97X-D/6-311++g(d,p) theory; Table S2: Theoretical and experimental vibrational wavenumbers (cm$^{-1}$) of NF-DMAP with PED using B3LYP functional; Table S3: Theoretical and experimental vibrational wavenumbers (cm$^{-1}$) of NF-DMAP with PED using wB97X-D level of theory; Table S4. Theoretical and experimental vibrational wavenumbers (cm$^{-1}$) of NF with PED using B3LYP level of theory; Table S5: Theoretical and experimental vibrational wavenumbers (cm$^{-1}$) of NF with PED using wB97X-D level of theory; Table S6: Geometrical parameters for intermolecular hydrogen bond in NF-DMAP; Table S7: Bond contributions to atomic net charges for NF-DMAP using QTAIM approach; Table S8: Second order perturbation theory analysis of Fock matrix in NBO basis for inter and intramolecular interactions within NF-DMAP using B3LYP functional; Table S9. Second order perturbation theory analysis of Fock matrix in NBO basis for inter and intramolecular interactions within NF-DMAP using wB97X-D functional; Table S10. Calculated local reactivity properties of atoms of NF-DMAP by Hirshfeld derived charges; Table S11. Calculated local reactivity properties of the atoms of NF using Hirshfeld derived charges.

**Author Contributions:** For research articles with several authors, a short paragraph specifying their individual contributions must be provided. The following statements should be used "conceptualization, E.K., P.T. and V.R.V.; methodology, E.K. and V.R.V.; software, X.X.; validation, E.K. and P.T.; formal analysis, E.K.; investigation, E.K. and V.R.V.; resources, E.K., P.T. and V.R.V.; data curation, E.K. and V.R.V.; writing—original draft preparation, E.K.; writing—review and editing, E.K., A.S., K.S., D.G., K.H.A., P.T. and V.R.V.; visualization, E.K.; supervision, P.T. and V.R.V.; project administration, P.T. and V.R.V.

**Funding:** This research was funded by University Grant Commission (UGC), India. V.R.V. thank Newton-Bhabha for Ph.D. placement award (2017) and Royal Society seed corn research grant (2018-19).

**Acknowledgments:** We thank A. P. Ayala, Universidade Federal do Ceará, Brazil, for executing Fourier Transform-Raman spectra of the salt. E. Khan and K. Srivastava are grateful to UGC, India, for providing the BSR-SRF meritorious fellowship. D. Gangopadhyay is grateful to SERB, DST, India, for providing the National Post-doctoral Fellowship (Project File Number: PDF/2016/003162). We acknowledge Central Facility for Computational Research (CFCR), University of Lucknow. V. R. Vangala thanks the Newton-Bhabha for Ph.D. placement award (2017) and Royal Society seed corn research grant (2018-19).

**Conflicts of Interest:** The authors declare no conflict of interest.

## References

1. Duggirala, N.K.; Perry, M.L.; Almarsson, Ö.; Zaworotko, M.J. Pharmaceutical cocrystals: Along the path to improved medicines. *Chem. Commun.* **2016**, *52*, 640–655. [CrossRef] [PubMed]
2. Grothe, E.; Meekes, H.; Vlieg, E.; ter Horst, J.H.; de Gelder, R. Solvates, salts, and cocrystals: A proposal for a feasible classification system. *Cryst. Growth Des.* **2016**, *16*, 3237–3243. [CrossRef]
3. Zhu, B.; Wang, J.R.; Zhang, Q.; Li, M.; Guo, C.; Ren, G.; Mei, X. Stable cocrystals and salts of the antineoplastic drug apatinib with improved solubility in aqueous solution. *Cryst. Growth Des.* **2018**, *18*, 4701–4714. [CrossRef]
4. Skorepova, E.; Bím, D.; Husak, M.; Klimes, J.; Chatziadi, A.; Ridvan, L.; Boleslavska, T.; Beranek, J.; Sebek, P.; Rulísek, L. Increase in solubility of poorly-ionizable pharmaceuticals by salt formation: A case of agomelatine sulfonates. *Cryst. Growth Des.* **2017**, *17*, 5283–5294. [CrossRef]
5. Brittain, H.G. Vibrational spectroscopic studies of cocrystals and salts. 1. The benzamide-benzoic acid system. *Cryst. Growth Des.* **2009**, *9*, 2492–2499. [CrossRef]

6. Brittain, H.G. Vibrational spectroscopic studies of cocrystals and salts. 2. The benzylamine-benzoic acid system. *Cryst. Growth Des.* **2009**, *9*, 3497–3503. [CrossRef]
7. Brittain, H.G. Vibrational spectroscopic studies of cocrystals and salts. 3. Cocrystal products formed by benzene carboxylic acids and their sodium salts. *Cryst. Growth Des.* **2010**, *10*, 1990–2003. [CrossRef]
8. Brittain, H.G. Vibrational spectroscopic studies of cocrystals and salts. 4. Cocrystal products formed by benzylamine, α-methylbenzylamine, and their chloride salts. *Cryst. Growth Des.* **2011**, *11*, 2500–2509. [CrossRef]
9. Aitipamula, S.; Vangala, V.R. X-ray crystallography and its role in understanding the physicochemical properties of pharmaceutical cocrystals. *J. Ind. Inst. Sci.* **2017**, *97*, 227–243. [CrossRef]
10. Pienaar, E.W.; Caira, M.R.; Lotter, A.P. Polymorphs of nitro-furantoin 1. Preparation and X-ray crystal-structures of two monohydrated forms of nitrofurantoin. *J. Crystallogr. Spectr. Res.* **1993**, *23*, 739–744. [CrossRef]
11. WHO Model List of Essential Medicines [WWW Document]. Available online: http://www.who.int/selectionmedicines/committees/expert/17/sixteenthadultlisten (accessed on 8 March 2017).
12. *Macrodantin Package Insert*; Procter & Gamble—US: Pineville, LA, USA, 2009.
13. Ebian, A.E.A.R.; Moustafa, R.M.A.; Abul-Enin, E.B. Nitrofurantoin. I. Effect of aging at different relative humidities and higher temperatures on the drug release and the physical properties of tablets. *Egypt J. Pharm. Sci.* **1985**, *26*, 287–300.
14. Ebian, A.E.A.R.; Fikrat, H.T.; Moustafa, R.M.A.; Abul-Enin, E.B. Nitrofurantoin. II. Correlation of *in vivo* bioavailability to in vitro dissolution of nitrofurantoin tablets aged at different relative humidities and elevated temperatures. *Egypt J. Pharm. Sci.* **1986**, *27*, 347–358.
15. Bertolasi, B.Y.V.; Gilli, P.; Ferretti, V.; Gilli, G. Structure and crystal packing of the antibacterial drug 1-[[(5-nitro-2-furanyl)methylene]amino}-2,4-imidazolidinedione (nitrofurantoin). *Acta Crystallogr. Sect. C* **1993**, *49*, 741–744. [CrossRef]
16. Marshall, P.V.; York, P. Crystallisation solvent induced solid- state and particulate modifications of nitrofurantoin. *Int. J. Pharm.* **1989**, *55*, 257–263. [CrossRef]
17. Otsuka, M.; Teraoka, R.; Matsuda, Y. Physicochemical stability of nitrofurantoin anhydrate and monohydrate under various temperature and humidity conditions. *Pharm. Res.* **1991**, *8*, 1066. [CrossRef]
18. Vangala, V.R.; Chow, P.S.; Tan, R.B.H. The solvates and salt of antibiotic agent, nitrofurantoin: Structural, thermochemical and desolvation studies. *CrystEngComm* **2013**, *15*, 878–889. [CrossRef]
19. Stephen Chan, H.C.; Kendrick, J.; Neumann, M.A.; Leusen, F.J.J. Towards abinitio screening of co-crystal formation through lattice energy calculations and crystalstructure prediction of nicotinamide, isonicotinamide, picolinamide and paracetamolmulti-component crystals. *CrystEngComm* **2013**, *15*, 3799. [CrossRef]
20. Srivastava, K.; Shukla, A.; Karthick, T.; Velaga, S.P.; Tandon, P.; Sinha, K.; Shimpi, M.R. Molecular structure, spectroscopic signature and reactivity analyses of paracetamol hydrochloride monohydrate salt using density functional theory calculations. *CrystEngComm* **2019**, *21*, 857–865. [CrossRef]
21. Sacchi, M.; Brewer, A.Y.; Jenkins, S.J.; Parker, J.E.; Friščić, T.; Clarke, S.M. Combined diffraction and density functional theory calculations of halogen-bonded cocrystal monolayers. *Langmuir* **2013**, *29*, 14903–14911. [CrossRef]
22. Shukla, A.; Khan, E.; Alsirawan, M.B.; Mandal, R.; Tandon, P.; Vangala, V.R. Spectroscopic (FT-IR, FT-Raman, 13C SS-NMR) and quantum chemical investigations to explore the structural insights of nitrofurantoin-4-hydroxybenzoic acid cocrystal. *New J. Chem.* **2019**, *43*, 7136–7149. [CrossRef]
23. Shukla, A.; Khan, E.; Srivastava, K.; Sinha, K.; Tandon, P.; Vangala, V.R. Study of molecular interactions and chemical reactivity of the nitrofurantoin-3-aminobenzoic acid cocrystal using quantum chemical and spectroscopic (IR, Raman, 13 C SS-NMR) approaches. *CrystEngComm* **2017**, *19*, 3921–3930. [CrossRef]
24. Khan, E.; Shukla, A.; Jhariya, A.N.; Tandon, P.; Vangala, V.R. Nitrofurantoin-melamine monohydrate (cocrystal hydrate): Probing the role of H-bonding on the structure and properties using quantum chemical calculations and vibrational spectroscopy. *Spectrochim. Acta Part. A Mol. Biomol. Spectrosc.* **2019**, *221*, 117170. [CrossRef] [PubMed]
25. Khan, E.; Shukla, A.; Jadav, N.; Telford, R.; Ayala, A.P.; Tandon, P.; Vangala, V.R. Study of molecular structure, chemical reactivity and H-bonding interactions in the cocrystal of nitrofurantoin with urea. *New J. Chem.* **2017**, *41*, 11069–11078. [CrossRef]
26. Hohenberg, P.; Kohn, W. Inhomogeneous electron gas. *Phys. Rev. B* **1964**, *136*, 864–871. [CrossRef]

27. Frisch, M.J.; Trucks, G.W.; Schlegel, H.B.; Scuseria, G.E.; Robb, M.A.; Cheeseman, J.R.; Scalmani, G.; Barone, V.; Mennucci, B.; Petersson, G.A.; et al. *Gaussian 09, Revision E.01*; Gaussian, Inc.: Wallingford, CT, USA, 2009.
28. Petersson, G.A.; Allaham, M.A. A complete basis set model chemistry. II. Open-shell systems and the total energies of the first-row atoms. *J. Chem. Phys.* **1991**, *94*, 6081–6091. [CrossRef]
29. Petersson, G.A.; Bennett, A.; Tensfeldt, T.G.; Allaham, M.A.; Shirley, W.A.; Mantzaris, J. A complete basis set model chemistry. I. The total energies of closed-shell atoms and hydrides of the first-row elements. *J. Chem. Phys.* **1988**, *89*, 2193–2218. [CrossRef]
30. Becke, A.D. Density-functional thermochemistry. III. The role of exact exchange. *J. Chem. Phys.* **1993**, *98*, 5648–5652. [CrossRef]
31. Lee, C.T.; Yang, W.T.; Parr, R.G. Development of the Colle-Salvetti correlation-energy formula into a functional of the electron density. *Phys. Rev. B* **1988**, *37*, 785–789. [CrossRef]
32. Parr, R.G.; Yang, W. *Density Functional Theory of Atoms and Molecules*; Oxford University Press: New York, NY, USA, 1989.
33. Chai, J.D.; Head-Gordon, M. Long-range corrected hybrid density functionals with damped atom-atom dispersion corrections. *Phys. Chem. Chem. Phys.* **2008**, *10*, 6615–6620. [CrossRef]
34. Pulay, P.; Fogarasi, G.; Pang, F.; Boggs, J.E. Systematic ab initio gradient calculation of molecular geometries, force constants, and dipole moment derivatives. *J. Am. Chem. Soc.* **1979**, *101*, 2550–2560. [CrossRef]
35. Martin, J.M.L.; Alsenoy, C.V. *Gar2ped*; University of Antwerp: Antwerpen, Belgium, 1995.
36. Frisch, A.; Nielson, A.B.; Holder, A.J. *Gaussview User Manual*; Gaussian, Inc.: Pittsburgh, PA, USA, 2000.
37. Bader, R.F.W.; Cheeseman, J.R. *AIMPAC: A Suite of Programs for the AIM Theory*; McMaster University: Hamilton, ON, Canada, 2000.
38. National Center for Biotechnology Information. Pubchem Compound Database [WWW Document]. Available online: https://pubchem.ncbi.nlm.nih.gov/compound/14284 (accessed on 15 January 2017).
39. Karasulu, B.; Götze, J.P.; Thiel, W. Assessment of Franck–Condon methods for computing vibrationally broadened UV–vis absorption spectra of flavin derivatives: Riboflavin, roseoflavin, and 5-thioflavin. *J. Chem. Theory Comput.* **2014**, *10*, 5549–5566. [CrossRef] [PubMed]
40. Scott, A.P.; Radom, L. Harmonic vibrational frequencies: An evaluation of hartree–fock, møller–plesset, quadratic configuration interaction, density functional theory, and semiempirical scale factors. *J. Phys. Chem.* **1996**, *100*, 16502–16513. [CrossRef]
41. Wong, M.W. Vibrational frequency prediction using density functional theory. *Chem. Phys. Lett.* **1996**, *256*, 391–399. [CrossRef]
42. Banwell, C.N.; McCash, E.M. *Fundamentals of Molecular Spectroscopy*; Mcgraw-Hill: New York, NY, USA, 1994.
43. Lambert, J.B. *Introduction to Organic Spectroscopy*; Macmillan: New York, NY, USA, 1987.
44. Delgado, M.C.R.; Hernandez, V.; Navarrete, J.T.L.; Tanaka, S.; Yamashita, Y. Electronic, optical, and vibrational properties of bridged dithienylethylene-based NLO chromophores. *J. Phys. Chem. B* **2004**, *108*, 2516–2526. [CrossRef]
45. Socrates, G. *Infrared and Raman Characteristic Group Frequencies: Tables and Charts*; John Wiley & Sons: Hoboken, NJ, USA, 2007.
46. Smith, B.C. *Infrared Spectral Interpretation: A Systematic Approach*; CRC Press: Boca Raton, FL, USA, 1999.
47. Koch, U.; Popelier, P.L.A. Characterization of C-H···O hydrogen bonds on the basis of the charge density. *J. Phys. Chem.* **1995**, *99*, 9747–9754. [CrossRef]
48. Rozas, I.; Alkorta, I.; Elguero, J. Behavior of ylides containing N, O, and C atoms as hydrogen bond acceptors. *J. Am. Chem. Soc.* **2000**, *122*, 11154–11161. [CrossRef]
49. Fukui, K. Role of frontier orbitals in chemical reactions. *Science* **1982**, *218*, 747. [CrossRef]
50. Khan, E.; Shukla, A.; Srivastava, A.; Shweta, P.; Tandon, P. Molecular structure, spectral analysis and hydrogen bonding analysis of ampicillin trihydrate: A combined DFT and AIM approach. *New J. Chem.* **2015**, *39*, 9800–9812. [CrossRef]
51. Chidangil, S.; Shukla, M.K.; Mishra, P.C. A molecular electrostatic potential mapping study of some fluoroquinolone anti-bacterial agents. *J. Mol. Model.* **1998**, *4*, 250–258. [CrossRef]

52. Luque, F.J.; Lopez, J.M.; Orozco, M. Perspective on electrostatic interactions of a solute with a continuum. A direct utilization of ab initio molecular potentials for the prevision of solvent effects. *Theor. Chem. Acc.* **2000**, *103*, 343–345. [CrossRef]
53. Parr, R.G.; Szentpaly, L.; Liu, S. Electrophilicity index. *J. Am. Chem. Soc.* **1999**, *121*, 1922. [CrossRef]

© 2019 by the authors. Licensee MDPI, Basel, Switzerland. This article is an open access article distributed under the terms and conditions of the Creative Commons Attribution (CC BY) license (http://creativecommons.org/licenses/by/4.0/).

Article

# Elucidation of the Crystal Structures and Dehydration Behaviors of Ondansetron Salts

Ryo Mizoguchi [1,2] and Hidehiro Uekusa [1,*]

[1] Department of Chemistry and Materials Science, Tokyo Institute of Technology, Ookayama 2-12-1, Meguro-ku, Tokyo 152-8551, Japan; uekusa@cms.titech.ac.jp

[2] Analytical Research Laboratories, Astellas Pharma Incorporation, 180, Ozumi, Yaizu-shi, Shizuoka 425-0072, Japan; ryo.mizoguchi@astellas.com

* Correspondence: uekusa@cms.titech.ac.jp; Tel.: +81-3-5734-3529

Received: 7 March 2019; Accepted: 25 March 2019; Published: 26 March 2019

**Abstract:** In drug development, it is extremely important to evaluate the solubility and stability of solid states and to immediately determine the potential for development. Salt screening is a standard and useful method for obtaining drug candidates with good solid state properties. Ondansetron is marketed as a hydrochloride dihydrate, and its dehydration behavior was previously reported to transition to an anhydrate via a hemihydrate as an intermediate by heating. Here, we synthesized ondansetron hydrobromide and hydroiodide and examined their dehydration behaviors. Single-crystal structure analysis confirmed that like ondansetron hydrochloride, ondansetron hydrobromide formed a dihydrate. Moreover, the crystal lattice parameters and hydrogen bonding networks were similar and isomorphic. While single-crystal structure analysis showed that ondansetron hydroiodide also formed a dihydrate, the crystal lattice parameters and hydrogen bonding networks were different to those of ondansetron hydrobromide and hydrochloride. Additionally, the dehydration behavior of ondansetron hydrobromide differed from that of the hydrochloride, with no hemihydrate intermediate forming from the hydrobromide, despite similar anhydrate structures. Given that it is difficult to predict how a crystal structure will form and the resulting physical properties, a large amount of data is needed for the rational design of salt optimization.

**Keywords:** crystal structure analysis; structure determination from powder diffraction data; salt optimization; hydrate; ondansetron; hygroscopicity; dehydration; physicochemical properties

## 1. Introduction

In drug discovery, salt screening is a general tool used to develop a solid form of a compound with improved physicochemical properties [1,2]. The counter acids or bases used in salt selection can be selected if the pka values are used as a reference [3,4]. Salt formation is expected to improve the compound's solubility, an important physicochemical property in drug development [5,6].

Salt formation is also expected to improve physicochemical properties other than solubility. For example, salt formation reportedly improved the hygroscopicity of the ethambutol dihydrochloride salt [7]. The improved salt was an oxalate. Penetration of water was inhibited because the packing efficiency of the oxalate crystals was improved compared to that in the dihydrochloride.

Salt optimization is selecting the counter ion, as the salt form has the most proper physicochemical properties as an active pharmaceutical ingredient (API). Although salt optimization is widely known in the pharmaceuticals field, few studies have reported the rational design for such optimization. Instead of a rational design, many screening methods have been developed [8,9]. Salt screening generally requires large amounts of drugs and time. Given that the amount of drugs and the time for research are limited in the early stages of drug discovery, new technology is needed to solve these problems. One answer may be to use rational design approaches for salt formation.

Rational design approaches require a working knowledge of the crystal structure of salts and aim to improve understanding of the relationship between the crystal structure and physicochemical properties. The final purpose of rational design is to control the physicochemical properties of salts by changing the salt-forming compounds. This requires predicting how changing the salt-forming compounds will affect the crystal structure and how the crystal structure represents the physicochemical properties.

Changing a solid state can lead to expected or unexpected changes in physiochemical properties. Unexpected changes to a solid state are sometimes observed in the manufacturing process or under various storage conditions [10,11]. For example, the hydrate form of a compound is sometimes observed unexpectedly as physical changes in hydrates occur due to changes in humidity or temperature [12,13]. Therefore, it is necessary to predict the occurrence of hydration or dehydration and how they change the physicochemical properties of a compound.

Understanding the mechanism of dehydration will enable the consideration of physicochemical properties before and after dehydration. A number of reports have examined the mechanism of dehydration of hydrates from crystal structures [14–16], including our previous reports [17,18]. For example, the mechanism of two-step dehydration of the dihydrate phase of lisinopril was clearly established from the crystal structures [19].

In this paper, we used ondansetron as a model compound. Ondansetron, also known as Zofran, was launched in the USA in 1991. Ondansetron is a competitive serotonin type 3 receptor antagonist and is effective for the treatment of nausea and vomiting caused by cytotoxic chemotherapeutic drugs [20]. The crystal structure of ondansetron hydrochloride dihydrate has been reported (CSD refcode: YILGAB) [21]. A previous study reported the dehydration reaction of ondansetron hydrochloride dihydrate [22], showing that it transformed into a hemihydrate intermediate prior to an anhydrate.

In the drug discovery, salt optimization will be conducted, and it would be interesting to evaluate whether the rare dehydration reaction of ondansetron hydrochloride dihydrate is observed in the formation of other salts. Here, we report the synthesis of ondansetron hydrobromide and ondansetron hydroiodide and evaluate their crystal structures and dehydration reactions.

## 2. Materials and Methods

*2.1. Materials*

Ondansetron hydrochloride dihydrate was purchased from Sigma Aldrich (St. Louis, MO, USA). Solvents for the synthesis reactions and the hydrobromide solution were purchased from Kanto Kagaku (Tokyo, Japan). Hydroiodide solution was purchased from Wako Junyaku (Tokyo, Japan).

Characterization was performed using samples pulverized with an agate mortar.

*2.2. Synthesis*

Ondansetron HBr salt was synthesized by dissolving ondansetron HCl dihydrate in methanol containing an equal molar quantity of sodium hydroxide. Subsequently, an equal molar quantity of hydrobromide in methanol solution was added to the methanol solution containing ondansetron and the solvent was removed under dry $N_2$ gas. After drying, acetone/water (9:1) was added and the sample was stirred overnight. The resulting powder was filtered and recrystallized using ethanol/$H_2O$ to obtain the bulk sample. In addition, single crystals were also obtained under a different recrystallization condition. Ondansetron HBr anhydrate B was obtained from a slurry suspension using ethyl acetate or acetonitrile instead of acetone/water.

Ondansetron HI salt was synthesized by dissolving ondansetron HCl dihydrate in methanol containing an equal molar quantity of sodium hydroxide. Subsequently, an equal molar quantity of hydroiodide in methanol solution was added to the methanol solution containing ondansetron and the solvent was removed under dry $N_2$ gas. 2-propanol/water (1:1) was added and bulk samples were recrystallized. In addition, single crystals were also obtained by recrystallization in 2-propanol/water (1:1). In addition, ondansetron HI anhydrate B was obtained using ethanol/water (9:1) instead of 2-propanol/water (1:1).

Samples were heated in a Fine Oven DF42 (Yamato Science, Tokyo, Japan).

### 2.3. Single Crystal Structure Analysis

All diffraction experiments were performed in a Rigaku XtaLAB P200 diffractometer using multi-layer mirror monochromated CuKα ($\lambda$ = 1.54184 Å) irradiation. The data were collected from a crystal at 93 K under a nitrogen stream. The initial structure was determined by the direct method using SHELXT [23], and intensity data were corrected using the Lorentz factor, polarization factor, and absorption correction. Non-hydrogen atoms were refined anisotropically. Hydrogen atoms, except for those of $H_2O$, were generated geometrically and refined using the riding model. Hydrogen atoms of $H_2O$ were located using difference Fourier synthesis, and refined isotropically. The obtained structure was refined using full-matrix least-squares refinement (w = $1/[\sigma^2(Fo^2) + (0.0999 \cdot P)^2 + 0.0818 \cdot P]$, P = $(Max(Fo^2,0) + 2Fc^2)/3)$) to minimize $\Sigma w(Fo^2-Fc^2)^2$.

The void volume was calculated with Mercury 3.10 (The Cambridge Crystallographic Data Centre (CCDC), Cambridge, United Kingdom) using default parameters.

### 2.4. Structure Determination from Powder Diffraction Data (SDPD)

The crystal structures of both the hemihydrate and anhydrate were determined from powder X-ray diffraction (PXRD) data measured using SPring-8 BL19B2. Crystal structure analysis was performed using the Powder Solve module of Materials Studio (BIOVIA, Tokyo, Japan). After selecting the peaks set, indexing was conducted in the X-CELL module to introduce the unit cell and appropriate space group [24]. The unit cell was refined using Pawley refinement and optimized.

The initial chemical structures of ondansetron and the water molecules were introduced using the Forcite module with COMPASS II as the force field [25]. The initial crystal structure was introduced using the Powder Solve module and the simulated-annealing approach, and optimized by Rietveld refinement [26]. Pareto optimization, a Rietveld refinement method that considers the energy of the structure calculated by a force field [27], was performed in the final optimization step.

### 2.5. Thermal Analysis: Differential Scanning Calorimetry (DSC)

Thermal analysis was performed using a TA Q2000 DSC instrument, which included a refrigerated cooling system (TA Instruments, New Castle, DE, USA). Temperature calibrations were performed using the indium metal standard supplied with the instrument. Samples were weighed (about 3 mg) in aluminum pans and analyzed from 25 °C to 300 °C at a heating rate of 10 °C/min using a similar empty pan as a reference. An inert atmosphere was maintained in the calorimeter by purging with nitrogen gas at a flow rate of 50 mL/min.

### 2.6. Thermal Analysis: Thermogravimetric Analysis (TGA)

Thermogravimetric analysis (TGA) was performed using a TA Q500 TGA instrument (TA Instruments). Approximately 4 mg of sample was loaded into a platinum pan and heated to 300 °C at a rate of 10 °C/min. Measurements were conducted under a nitrogen purge at a flow rate of 100 mL/min. Temperature calibration was performed using standard nickel.

### 2.7. Powder X-ray Diffraction

PXRD measurements were performed on TTR II (Rigaku, Tokyo, Japan) with Cu Kα radiation at 1.54184 Å, a voltage of 50 kV and current of 300 mA. Data were collected at a scan rate of 4°/min over a 2θ range from 2.5° to 40°. In variable temperature PXRD measurements (VT-XRD), simultaneous measurement of powder X-ray diffraction data and differential scanning calorimetry (XRD-DSC) was performed on a SmartLab system (Rigaku, Tokyo, Japan) using Cu Kα radiation at 1.54184 Å, a voltage of 45 kV and current of 200 mA, with a DSC attachment and a D/Tex Ultra detector. Samples were weighed (1.5–2.5 mg) in aluminum pans and analyzed at a heating rate of 2 °C/min using a similar but empty pan as a reference.

For the synchrotron X-ray measurements using SPring-8, the powder samples were enclosed in a 0.3-mm Lindemann glass capillary. The X-ray powder diffraction data were collected using SPring-8 BL19B2, which was equipped with a high-resolution type Debye–Scherrer camera and a curved imaging-plate detector [28]. The wavelength was set at 1.0000 Å. For the variable temperature measurements, it took one minute to set each temperature. Each temperature was then maintained for 4 min to ensure that equilibrium was reached before the measurements were taken. Data were collected for 5 min. During data collection, the sample was maintained at the set temperature and rotated at 1 r/min to reduce potential preferential orientation effects.

### 2.8. Water Vapor Sorption and Desorption Studies

Dynamic vapor sorption experiments were performed using VTI SGA 100 (VTI Corporation, Hialeah, FL, USA). Samples (about 10 mg) were studied over a selected humidity range (absorption process: from 5% relative humidity (RH) to 95% RH; desorption process: from 95% RH to 5% RH) at 25 °C. For each humidity step, the equilibration was set to dm/dt 0.03%/min on a 5-min time frame (maximum hold time 180 min).

## 3. Results

### 3.1. Elucidation of the Crystal Structure of Ondansetron Salts

Crystal graphic parameters of the ondansetron HCl salt, HBr salt and HI salt are summarized in Table 1. A list of interactions observed in each crystal are shown in the Supporting Information (Table S1).

The results of single crystal structure analysis of ondansetron salts dihydrate are shown in Figure 1.

Single crystal structure analysis confirmed that ondansetron HBr formed a dihydrate. Ondansetron HBr had a similar crystal structure to ondansetron HCl [21]. These crystals structures were isomorphous, as evidenced by the unit cell parameters and hydrogen bonding networks. In particular, the water molecules were in direct contact with the imidazole cation, and the bromide anion interacted with the water molecules. Moreover, there were other water molecules between the two bromide anions. A π-π stacking interaction and CH-π were observed between the tricyclic groups of ondansetron (Figure 1E), which we speculate act to stabilize the crystal structure.

The crystal structure of ondansetron HI also confirmed that it formed a dihydrate. In the dihydrate structure, the imidazole cation interacted with the water molecule, but the iodide anion did not directly interact with the imidazole cation. The diamond-shaped hydrogen bonding network observed in the crystal structures of HCl and HBr salts was not formed. Instead, a hydrogen bond formed between the water molecules. In contrast, π-π stacking interaction and CH-π interactions between the tricyclic groups were maintained (Figure 1F). However, a new stacking interaction was observed between the imidazole rings in the HI salt dihydrate (Figure 1D).

Comparison of unit cell parameters indicated that the HCl salt and HBr salt were isomorphous. In contrast, the HI dihydrate salt did not exhibit isomorphism because the unit cell of the HI salt was markedly different such that the space group was P-1, as shown in Table 1.

Table 1. Crystal graphic parameters of the ondansetron HCl salt, HBr salt and HI salt.

| | HCl salt (YILGAB [21]) | HBr salt dihydrate | HBr salt anhydrate A | HBr salt anhydrate B | HI salt dihydrate | HI salt anhydrate B |
|---|---|---|---|---|---|---|
| Chemical formula | $C_{18}H_{20}ClN_3O \cdot 2H_2O$ | $C_{18}H_{20}BrN_3O \cdot 2H_2O$ | $C_{18}H_{20}BrN_3O$ | $C_{18}H_{20}BrN_3O$ | $C_{18}H_{20}IN_3O \cdot 2H_2O$ | $C_{18}H_{20}IN_3O$ |
| Crystal system | Monoclinic | Monoclinic | Monoclinic | Monoclinic | Triclinic | Monoclinic |
| Space group | $P2_1/c$ | $P2_1/c$ | $P2_1/c$ | $P2_1/c$ | $P\text{-}1$ | $P2_1/c$ |
| Temp. (K) | 298 | 93 | 413 | 298 | 93 | 93 |
| $a$ (Å) | 15.082 (3) | 15.19912 (12) | 14.7114 (23) | 13.1939 (22) | 7.69854 (14) | 13.4834 (4) |
| $b$ (Å) | 9.741 (3) | 9.66181 (8) | 9.8069 (16) | 8.7052 (14) | 8.42520 (15) | 8.44016 (19) |
| $c$ (Å) | 12.734 (3) | 12.69645 (12) | 13.1735 (21) | 15.1708 (26) | 15.6102 (3) | 15.7830 (4) |
| $\alpha$ (°) | 90.0000 | 90.0000 | 90.0000 | 90.0000 | 90.2705 (14) | 90.0000 |
| $\beta$ (°) | 100.83 (1) | 100.6910 (8) | 114.1036 (8) | 104.1480 (3) | 96.0841 (15) | 111.035 (3) |
| $\gamma$ (°) | 90.0000 | 90.0000 | 90.0000 | 90.0000 | 107.9273 (17) | 90.0000 |
| $V$ (Å$^3$) | 1837.5 (8) | 1832.12 (3) | 1734.9 (5) | 1689.6 (5) | 957.17 (3) | 1676.45(8) |
| $Z$ | 4 | 4 | 4 | 4 | 2 | 4 |
| Radiation type | Cu Kα | Cu Kα | Synchrotron(λ = 1.0000) | Synchrotron(λ = 1.0000) | Cu Kα | Cu Kα |
| $R[F^2>2\sigma(F^2)]$ | 0.07 | 0.061 | - | - | 0.026 | 0.041 |
| $wR(F^2)$ | - | 0.196 | - | - | 0.068 | 0.107 |
| No. of reflections | 3219- | 3590 | - | - | 3689 | 3294 |
| No. of parameters | - | 240 | - | - | 240 | 210 |
| $R_p$ (%) | - | - | 8.56 | 7.18 | - | - |
| $R_{wp}$ (%) | - | - | 14.35 | 12.11 | - | - |
| Sample type | Single crystal (YILGAB) | Single crystal | Powder | Powder | Single crystal | Single crystal |
| CCDC No. | | 1893917 | 1893918 | 1893919 | 1893920 | 1893921 |

-: not obtained by a reference.

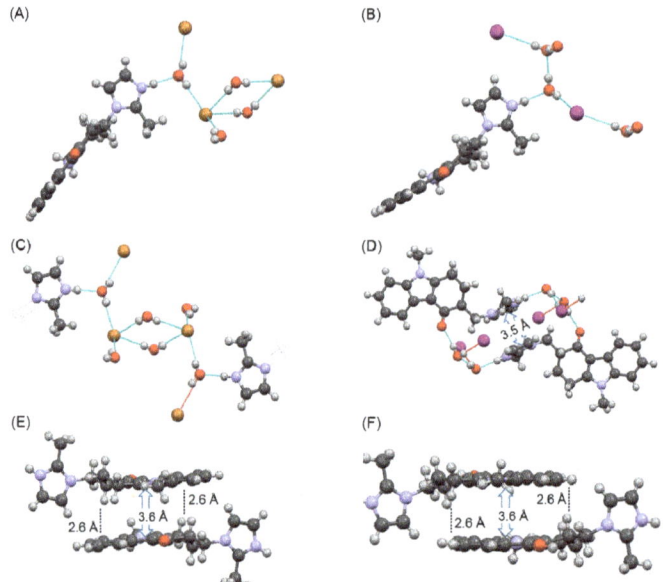

**Figure 1.** Results of single crystal structure analysis of the hydrogen bonding network in the (**A,C**) HBr salt, and (**B,D**) HI salt, and π-π stacking interactions in the (**E**) HBr salt and (**D,F**) HI salt.

### 3.2. Dehydration Behaviors of Ondansetron HBr and HI Dihydrates

Dehydration properties were examined using the crystalline powder of ondansetron HBr dihydrate and HI dihydrate. The results of thermal analysis of HBr dihydrate are shown in Figure 2A.

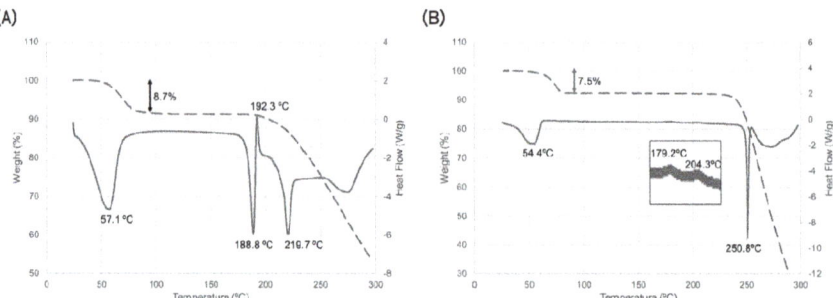

**Figure 2.** Thermal behavior of ondansetron salts. (**A**) HBr dihydrate and (**B**) HI dihydrate. Solid line indicates the differential scanning calorimetry (DSC) curve, dashed line indicates the thermogravimetric (TG) curve. Temperature on the DSC curve indicates the peak top temperature.

A broad endothermic peak associated with decreasing weight was observed at around room temperature and was hypothesized to correspond to the dehydration of water molecules. Given that the magnitude of weight decrease was equal to the theoretical weight of the water of a dihydrate, this suggested that the dehydration reaction was complete, and the thermal behavior of an anhydrate should be observed at more than 100 °C. An endothermic peak and an exothermic peak were observed around 190° C. These peaks indicated the melting temperature of ondansetron HBr anhydrate (anhydrate A) and crystallization of another form, respectively. Finally, an endothermic peak was observed around 220 °C. This peak indicated the melting temperature of another ondansetron HBr anhydrate (anhydrate B).

Dehydration properties were examined using the crystalline powder of ondansetron HI dihydrate. The results of the thermal analysis are shown in Figure 2B. The endothermic peak associated with the decrease in weight was observed at around room temperature. The magnitude of the weight decrease was equal to the theoretical weight of the water in ondansetron HI dihydrate. This indicates that dehydration of the dihydrate form occurred at relatively low temperature. The endothermic peak at around 250 °C indicates the melting temperature of the anhydrate appeared after dehydration.

VT-XRD experiments were performed to gain a deeper understanding of the relationship between the thermal behavior and crystal form transition.

The XRD pattern was different from that of the dihydrate from the start of the experiment, as shown in Figure 3A, indicating that dehydration had already begun when the sample was being dried under dry $N_2$ gas. The blue pattern in Figure 3A corresponds to that of an anhydrate because dehydration was observed in TGA (Figure 2A). In addition, the transition to another crystal form occurred at around 190 °C. This temperature corresponds to the thermal event observed in DSC (Figure 2A). In summary, ondansetron HBr dihydrate transformed to anhydrate A and then to an anhydrate B with heating.

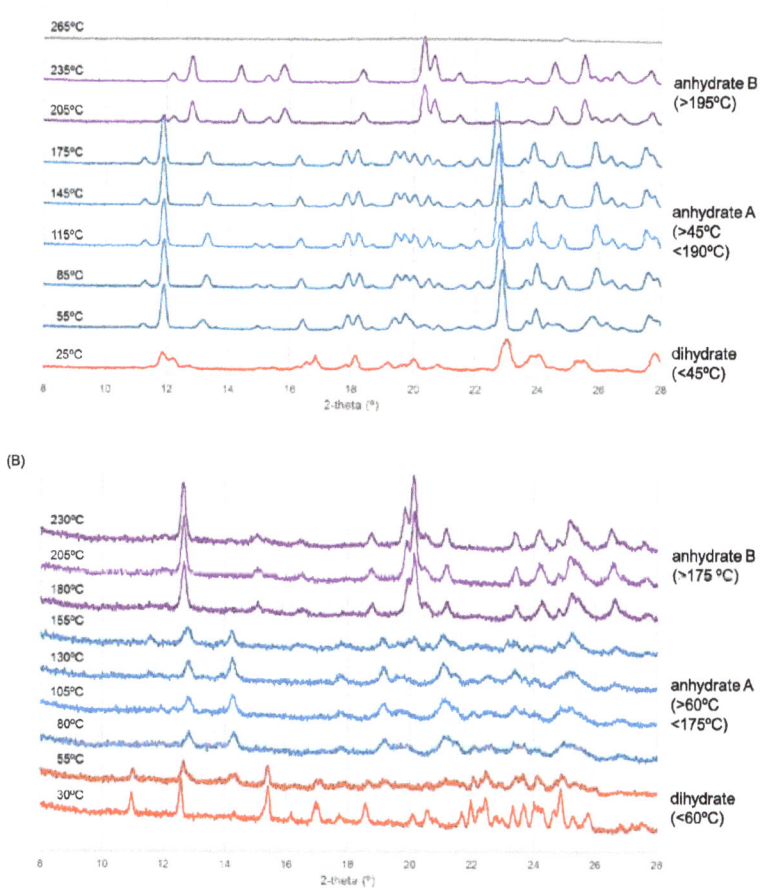

**Figure 3.** Variable temperature (VT)-X-ray diffraction scanning results for ondansetron (**A**) HBr dihydrate and (**B**) HI dihydrate. Patterns of the dihydrate (red), anhydrate A (blue), anhydrate B (purple) are shown.

As shown in Figure 3B, HI dihydrate transformed to anhydrate A at around 60 °C. The blue pattern in Figure 3B corresponds to that of an anhydrate because dehydration was observed in TGA (Figure 2B). Moreover, transition to another crystal form occurred around 170 °C. The temperature and transition may correspond to the minute exothermal peak observed in DSC (Figure 2B). In summary, ondansetron HI dihydrate likewise transformed to anhydrate A and then to an anhydrate B with heating.

### 3.3. Isolation of HBr and HI Anhydrates

We attempted to isolate the anhydrate A of the HBr salt and HI salt with heating. For both salts, the dihydrate form was heated to 80 °C for 30 min to form the anhydrate A. The samples were then left under ambient conditions for varying periods of time, after which PXRD patterns were measured over time.

As shown in Figure 4A, although the XRD pattern measured immediately after heating was different from the initial pattern, it reverted back to the initial pattern after 30 minutes at room temperature. Therefore, it was difficult to isolate the anhydrate A of an HBr salt at room environment. In contrast, as shown in Figure 4B, the anhydrate A of the HI salt was maintained for 4 h after heating. Heating caused the test materials to change slightly from white to pale brownish yellow. Therefore, the anhydrate A of an HI salt can be isolated at room environment.

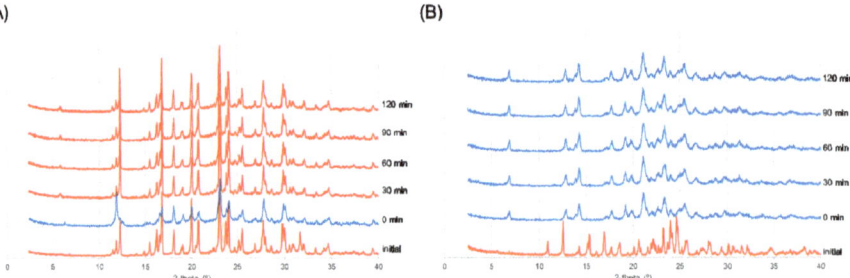

**Figure 4.** Reversibility after heating was confirmed from X-ray diffraction patterns. Red indicates the dihydrate diffraction pattern, and blue indicates the anhydrate diffraction pattern. (**A**) HBr salt and (**B**) HI salt.

### 3.4. Crystal Structure Analysis of the Anhydrates of HBr and HI Salts

To examine the mechanism governing the dehydration of ondansetron HBr dihydrate and HI dihydrate, we conducted crystal structure analysis of the PXRD patterns of anhydrate A and B. For crystal structure analysis of PXRD, XRD with heating was conducted using SPring-8 BL19B2. The results are shown in Figures S5 and S6.

Similar pattern changes to those shown in Figure 3 were observed in the SPring-8 BL19B2 experiments. For the HBr salt, crystal structure analysis was performed using the PXRD pattern at 140 or 25 °C cooling (in Figure S5) for anhydrate A and B, respectively. We attempted to isolate anhydrate B. Isolation was difficult with heating at around 200 °C due to the occurrence of melting and degradation. We therefore expect that strict temperature control is needed. However, isolation was possible from a slurry suspension of the dihydrate in ethyl acetate at 50 °C. While anhydrate B of the HBr salt could be isolated, single crystals could not be obtained. For the HI salt, although a crystal form transition was observed with heating, XRD patterns of anhydrate A and B were unclear, making it difficult to resolve the crystal structure using PXRD. Nevertheless, anhydrate B of the HI salt was isolated, as was the single crystal form (see the experimental section).

The crystal structures of two anhydrates of the HBr salt and one anhydrate of the HI salt were obtained. The structures of anhydrate A and B of the HBr salt and anhydrate B of the HI salt are shown in Figure 5.

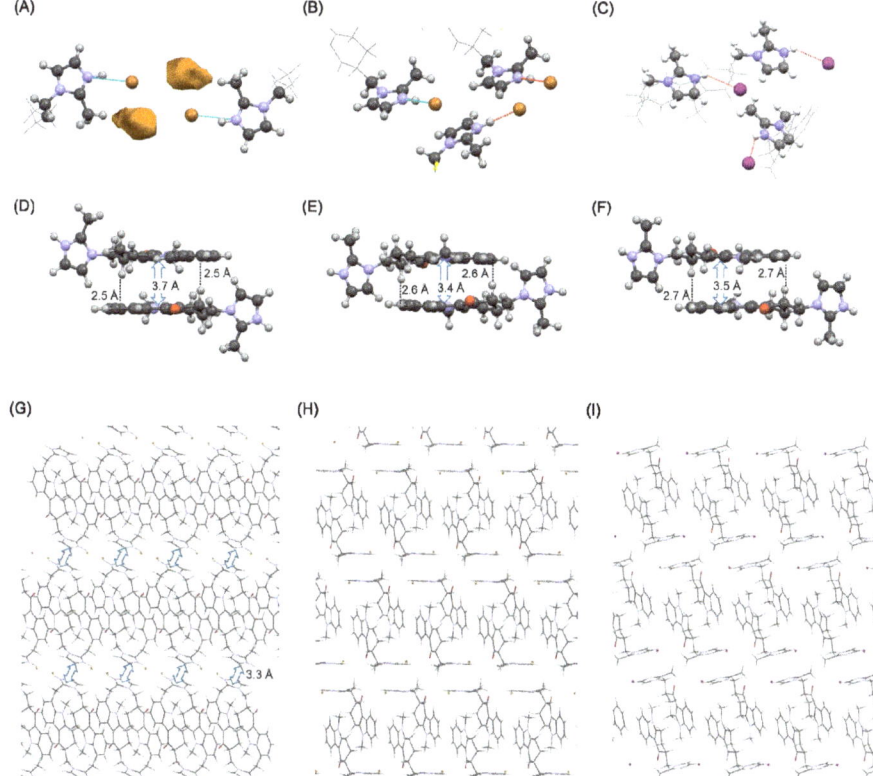

**Figure 5.** Comparison of anhydrate crystal forms. (**A–C**) Hydrogen bonding network, (**D–F**) π-π stacking interaction, and (**G–I**) B axis-projected view. (A,D,G) Anhydrate A of HBr, (B,E,H) anhydrate B of HBr, and (C,F,I) anhydrate B of HI.

According to the crystal structure of anhydrate A of HBr, the tricyclic groups of ondansetron were oriented in anti-parallel. This motif was also observed in the dihydrate, in which the imidazole cation did not directly interact with the bromide anion and the water molecule was intercepted. However, in anhydrate A, the absence of the water molecule indicates that a cation was in direct interaction with an anion. The space produced by dehydration was filled by translation of the imidazole ring, and new stacking interactions between the imidazole rings were formed (Figure 5G). These changes in interactions are similar to those observed in dehydration of the HCl salt.

In anhydrate B of HBr, the dimer structure formed by the tricyclic groups was likewise maintained. Direct interaction between the imidazole cation and bromide anion was also observed. Stacking interaction between the imidazole rings however was not observed. The void structure was likewise not observed, which may underlie the relative stability of anhydrate B at room environment. As anhydrate B also transformed to a dihydrate under higher humidity conditions, the dihydrate was the most stable form in high water activity environments.

In anhydrate B of HI, an ionic bond was formed between the imidazole cation and iodide anion. The stacking interaction and CH-π that were observed in all crystal phases of ondansetron were also maintained in anhydrate B. The stacking interaction between the imidazole rings however was not observed. The space group was $P2_1/c$, differing from that of the dihydrate.

## 3.5. Hygroscopicity

The dehydration tendency of ondansetron HBr and HI dihydrates at varying levels of RH was studied and the results are shown in Figure 6.

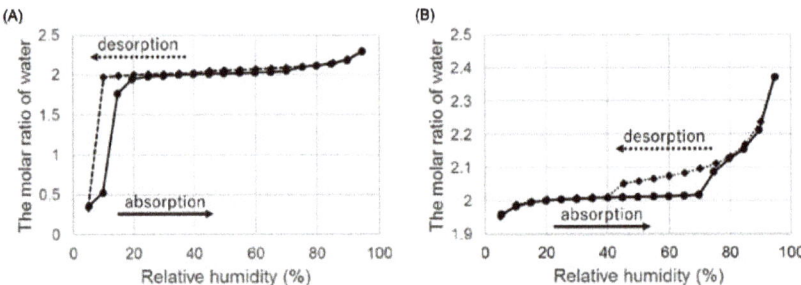

**Figure 6.** Hygroscopicity of ondansetron salts dihydrate at 25 °C. (**A**) Hydrobromide dihydrate and (**B**) hydroiodide dihydrate.

When we examined the weight change from 5% to 95% RH, there was a decrease in weight at low RH (Figure 6A). In the absorption process, the dihydrate of HBr was reformed and hydration was completed at 20% RH. These hygroscopicity profiles are consistent with the fact that it was difficult to isolate anhydrate A of HBr at room environment.

The dihydrate of HI showed the hygroscopicity at high RH, but no decrease in weight was observed at low humidity. The dihydrate was stable at low RH.

Hygroscopicity was studied to evaluate the physicochemical properties of the anhydrates (Figure 7).

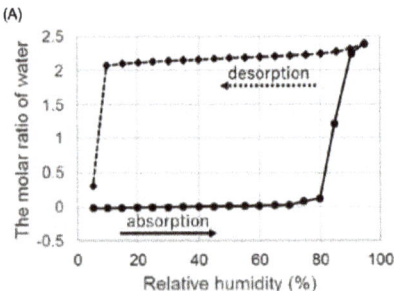

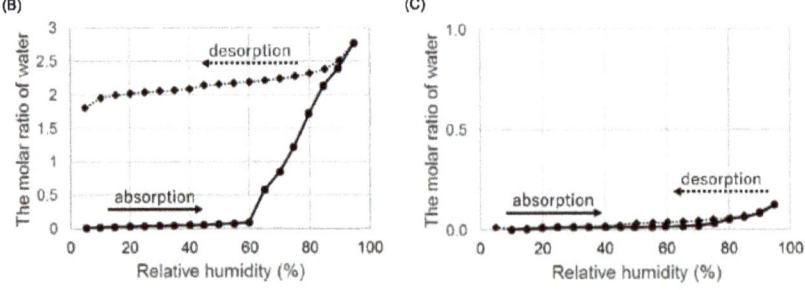

**Figure 7.** Hygroscopicity of ondansetron salts anhydrates at 25 °C. (**A**) Hydrobromide anhydrate B, (**B**) hydroiodide anhydrate A, and (**C**) hydroiodide anhydrate B.

When we examined the weight change from 5% to 95% RH, an anhydrate B of HBr absorbed water at around 85% RH and the increase of weight was equal to the theoretical weight of the dihydrate. The increased weight was maintained until RH reached below 10% in the desorption process, and the weight decreased from around 5% RH. This dehydration profile was the same as that of the dihydrate, indicating that an anhydrate B transformed to a dihydrate at high relative humidity.

Next, we evaluated the hygroscopicity of an anhydrate A of HI. The results are shown in Figure 7B. An anhydrate A absorbed water at around 65% RH, and the increase in weight was approximately equal to the weight of two water molecules. The increase in weight did not decrease at low RH. Confirmation of the crystal form using PXRD indicated that the anhydrate transformed back to a dihydrate. In summary, an anhydrate A obtained by heating a dihydrate transformed back to the initial dihydrate under high humidity conditions.

Finally, we evaluated the hygroscopicity of an anhydrate B of HI. The results are shown in Figure 7C. Although the anhydrate B of ondansetron hydroiodide did not show hygroscopicity, the anhydrate A did and transformed into a dihydrate. The low hygroscopicity of anhydrate B suggests that it may be suitable as a drug substance.

## 4. Discussion

The structures of ondansetron HBr dihydrate and HCl dihydrate were isomorphous, suggesting that HBr dihydrate should show a similar dehydration profile to HCl dihydrate forms. In particular, we evaluated whether or not a hemihydrate intermediate was formed through the dehydration process of HBr dihydrate. TGA produced a derivation curve with only one peak (ca. 60 °C), indicating that the dehydration progressed with one-step (Figure 8A). Whether or not an intermediate was formed was confirmed by monitoring the weight change at different temperatures (Figure 8B). The magnitude of weight change was equivalent to the weight of the water in ondansetron HBr dihydrate (about 9%). This indicates that, contrary to our expectation, the dihydrate of the HBr salt directly transformed to an anhydrate without forming a hemihydrate intermediate, as observed in the dehydration of the HCl salt.

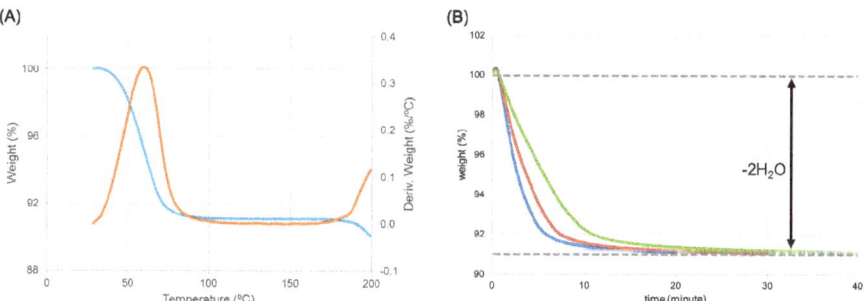

**Figure 8.** Identification of the intermediate using thermogravimetric analysis. (**A**) The thermogravimetric analysis (TGA) curve (blue) was overwritten by the derivation of the TGA curve (orange). (**B**) Change in weight (%) at different temperatures: 40 °C (green), 45 °C (red), and 50 °C (blue). The magnitude of weight loss (about 9%) was equivalent to the weight of water in the dihydrate.

To determine why no intermediate was observed in the dehydration of ondansetron HBr dihydrate, we considered the isomorphic relationship between ondansetron HCl and HBr dihydrates. Some related parameters of the anhydrates are shown in Table 2.

The ion radius of a bromide anion is bigger than that of a chloride anion, suggesting that the distance between two bromide anions is longer than that between two chloride anions. The distance between two anions should get longer considering the difference in ion radius, so the distance between bromide anions was relatively short comparing between chloride anions. As a result, the void volume was markedly different and the unit cell volume of the HBr salt was smaller than that of the HCl salt. The void space should be occupied by the water molecule in a hemihydrate. The small void volume of the HBr salt suggests that it is unable to retain a water molecule due to the lack of space. This may be a fundamental reason as to why no intermediate of HBr salt formed based on crystal structure analysis. Moreover, comparison of the density shows that the HBr salt has high packing efficiency, making it relatively stable. The relative instability of the anhydride of the HCl salt may explain the need to form an intermediate in the transition process.

Table 2. Comparison of parameters of the anhydrates of hydrochloride and hydrobromide.

|  | Hydrochloride [22] | Hydrobromide |
|---|---|---|
| Distance between anions (Å) | 4.755 | 4.767 |
| Ion radius (Å) [29] | 1.81 | 1.96 |
| Void volume (Å$^3$) | 6.1 | 1.3 |
| Unit cell volume (Å$^3$) | 1751 | 1735 |
| Density (g/cm$^3$) | 1.25 | 1.43 |

Both unit cells were determined at the same temperature (413 K).

When we examined the relationship between the crystal structures of each salt dihydrate, we found that the HI salt did not exhibit an isomorphic structure (Table 1 and Figure 1). The ion radius of the iodide anion (2.20 Å) is bigger than those of chloride and bromide anions [29]. Although it is possible for the crystal to increase in size and to rearrange, the larger anion may not fit into an isomorphic structure. Alternatively, studies have reported hydrated anions, represented as $[X(H_2O)]^-$ [30,31]. We observed a $[Cl_2(H_2O)_4]^{2-}$-like anion and a $[Br_2(H_2O)_4]^{2-}$-like anion in the crystal structure of ondansetron HCl and HBr dihydrate, respectively. Although a kind of $[Br_2(H_2O)_4]^{2-}$-like anion has not been reported, yet, similar types of hydrated anions have been reported. However, hydrated iodide anions reportedly form a different network. Chloride and bromide anions form cyclic clusters, while iodide anions form open chains [31]. We observed a chain cluster in the ondansetron HI dihydrate. The characteristics of the hydrated anions are likely related with the isomorphism of the corresponding dihydrates.

It is important to determine the most suitable property for each salt and hydrate of ondansetron from the perspective of drug development. Good solid physicochemical properties are required of the drug substance form. Stability against humidity is also important. Dihydrates of ondansetron HCl and HBr salts were unstable under low RH conditions, leading to the formation of an anhydrate or a hemihydrate. Anhydrate A of HCl and HBr salts were also unstable, and were even difficult to isolate at room environment. Anhydrate B of the HBr salt was unstable under high RH conditions, leading to the formation of a dihydrate. However, the dihydrate and anhydrate B of the HI salt were stable against humidity, indicating that both forms are feasible for drug development. As iodine has a characteristic pharmacokinetics profile in that it accumulates in the thyroid, iodine-containing salts tend to be avoided as drug substances. Therefore, salt selection requires the consideration of various factors.

Synthesis of the ondansetron HBr salt and HI salt was successful. A schematic of the dehydration of ondansetron salts is shown in Figure 9.

Ondansetron HCl, HBr and HI salts formed dihydrates. The structures of ondansetron HCl and HBr dihydrates were isomorphous, while ondansetron HI dihydrate had a different structure. Ondansetron HBr did not exhibit the two-stage dehydration observed in the reaction of ondansetron HCl dihydrate despite the isomorphic crystal structures of the dihydrates, indicating that isomorphic polymorphs do not always show the same physicochemical properties. The ondansetron HI salt dihydrate was stable at varying levels of humidity, and was more stable than the HCl and HBr dihydrates. Their hydrophilic properties suggest that halogen anions likely form hydrogen bonds with water molecules. Exchange of the halogen anion contributed to and was effective for improving the hygroscopicity.

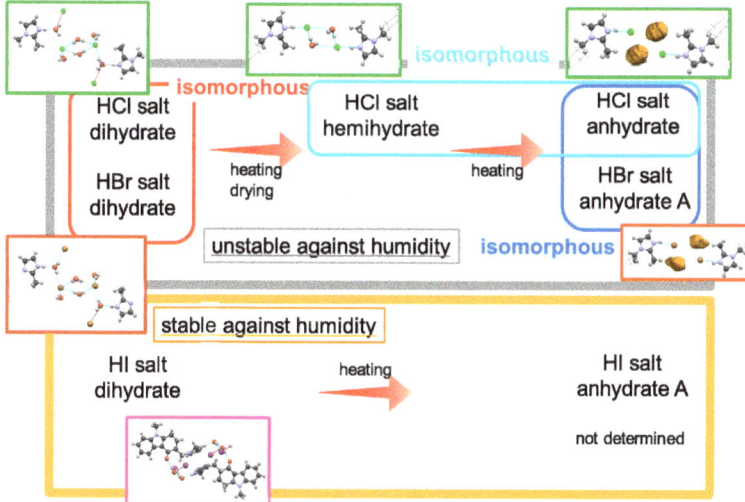

**Figure 9.** Schematic representation of the dehydration tendency of ondansetron salts.

Ondansetron HBr anhydrate B and HI anhydrate A and B were stable at room environment although HCl anhydrate salt was unstable. Unfortunately, ondansetron HBr anhydrate B and HI anhydrate A were unstable at high humidity, and were therefore unsuitable as development forms. The crystal structures of both anhydrates were markedly different from that of the anhydrate of HCl salt, which was unstable at room environment and was difficult to isolate. These results also indicate that exchanging the halogen anion contributed to the changing of the crystal structure. Given that it is difficult to predict how a crystal structure will form and the resulting physical properties, a large amount of data is needed for the rational design of salt optimization.

## 5. Conclusions

Ondansetron HCl, HBr and HI salts were synthesized and formed dihydrates. The structures of ondansetron HCl and HBr dihydrates were isomorphous, while ondansetron HI dihydrate showed a different structure. Ondansetron HBr dihydrate did not exhibit the two-stage dehydration observed in the reaction of ondansetron HCl dihydrate despite the isomorphic crystal structures of the dihydrate forms, indicating that isomorphic polymorphs do not always show the same physicochemical properties. The ondansetron HI dihydrate salt was stable across varying levels of humidity, suggesting that it was more stable than HCl and HBr dihydrate salts. These results also indicate that the changing of halogen anions contributed to changing the crystal structure. Given that it is difficult to predict how a crystal structure will form and the resulting physical properties, a large among of data is needed for the rational design of salt optimization.

**Supplementary Materials:** The following are available online at http://www.mdpi.com/2073-4352/9/3/180/s1: Figure S1. Numbering of atoms for each crystal form; Figure S2. XRD patterns of ondansetron salts; Figure S3. DSC curves of ondansetron salts; Figure S4. TGA curves of ondansetron salts; Figure S5. XRD with heating measurement of HBr dihydrate using SPring-8 BL19B2; Figure S6. XRD with heating measurement of HI dihydrate using SPring-8 BL19B2; and Table S1. List of interactions observed in each crystal structure.

**Author Contributions:** The manuscript was written through contributions from all authors. All authors have approved the final version of manuscript.

**Funding:** A part of this work was supported by JSPS KAKENHI Grant Number JP17K05745 and JP18H04504.

**Acknowledgments:** The synchrotron radiation experiments were performed using the BL19B2 of SPring-8 with the approval of the Japan Synchrotron Radiation Research Institute (JASRI) (Proposal No. 2016B1838). The authors also appreciate Yukihito Sugano for his assistance with X-ray crystal structure analysis and deep discussion.

**Conflicts of Interest:** The authors declare no conflict of interest.

## References

1. Tong, W.-Q.T.; Whitesell, G. In Situ Salt Screening-A Useful Technique for Discovery Support and Preformulation Studies. *Pharm. Dev. Technol.* **1998**, *3*, 215–223. [CrossRef]
2. Morissette, S.L.; Almarsson, Ö.; Peterson, M.L.; Remenar, J.F.; Read, M.J.; Lemmo, A.V.; Ellis, S.; Cima, M.J.; Gardner, C.R. High-throughput crystallization: Polymorphs, salts, co-crystals and solvates of pharmaceutical solids. *Adv. Drug Deliv. Rev.* **2004**, *56*, 275–300. [CrossRef] [PubMed]
3. He, Y.; Orton, E.; Yang, D. The Selection of a Pharmaceutical Salt—The Effect of the Acidity of the Counterion on Its Solubility and Potential Biopharmaceutical Performance. *J. Pharm. Sci.* **2018**, *107*, 419–425. [CrossRef] [PubMed]
4. Kim, H.; Gao, J.; Burgess, D.J. Evaluation of solvent effects on protonation using NMR spectroscopy: Implication in salt formation. *Int. J. Pharm.* **2009**, *377*, 105–111. [CrossRef]
5. Bayliss, M.K.; Butler, J.; Feldman, P.L.; Green, D.V.S.; Leeson, P.D.; Palovich, M.R.; Taylor, A.J. Quality guidelines for oral drug candidates: Dose, solubility and lipophilicity. *Drug Discovery Today* **2016**, *21*, 1719–1727. [CrossRef] [PubMed]
6. Serajuddin, A.T.M. Salt formation to improve drug solubility. *Adv. Drug Deliv. Rev.* **2007**, *59*, 603–616. [CrossRef]
7. Diniz, L.F.; Carvalho, P.S.; de Melo, C.C.; Ellena, J. Reducing the Hygroscopicity of the Anti-Tuberculosis Drug (S,S)-Ethambutol Using Multicomponent Crystal Forms. *Cryst. Growth Des.* **2017**, *17*, 2622–2630. [CrossRef]
8. Morissette, S.L.; Soukasene, S.; Levinson, D.; Cima, M.J.; Almarsson, Ö. Elucidation of crystal form diversity of the HIV protease inhibitor ritonavir by high-throughput crystallization. *Proc. Natl. Acad. Sci. USA* **2003**, *100*, 2180–2184. [CrossRef]
9. Remenar, J.F.; MacPhee, J.M.; Larson, B.K.; Tyagi, V.A.; Ho, J.H.; McIlroy, D.A.; Hickey, M.B.; Shaw, P.B.; Almarsson, Ö. Salt Selection and Simultaneous Polymorphism Assessment via High-Throughput Crystallization: The Case of Sertraline. *Org. Process. Res. Dev.* **2003**, *7*, 990–996. [CrossRef]
10. Bauer, J.; Spanton, S.; Henry, R.; Quick, J.; Dziki, W.; Porter, W.; Morris, J. Ritonavir: An Extraordinary Example of Conformational Polymorphism. *Pharm. Res.* **2001**, *18*, 859–866. [CrossRef]
11. Thakral, N.K.; Kelly, R.C. Salt disproportionation: A material science perspective. *Int. J. Pharm.* **2017**, *520*, 228–240. [CrossRef] [PubMed]
12. Tieger, E.; Kiss, V.; Pokol, G.; Finta, Z.; Dusek, M.; Rohlicek, J.; Skorepova, E.; Brazda, P. Studies on the crystal structure and arrangement of water in sitagliptin L-tartrate hydrates. *Crystengcomm* **2016**, *18*, 3819–3831. [CrossRef]
13. Grobelny, P.; Mukherjee, A.; Desiraju, G.R. Polymorphs and hydrates of Etoricoxib, a selective COX-2 inhibitor. *Crystengcomm* **2012**, *14*, 5785–5794. [CrossRef]
14. Bond, A.D.; Cornett, C.; Larsen, F.H.; Qu, H.; Raijada, D.; Rantanen, J. Structural basis for the transformation pathways of the sodium naproxen anhydrate–hydrate system. *Iucrj* **2014**, *1*, 328–337. [CrossRef] [PubMed]
15. Noguchi, S.; Miura, K.; Fujiki, S.; Iwao, Y.; Itai, S. Clarithromycin form I determined by synchrotron X-ray powder diffraction. *Acta Crystallogr. Sect. C* **2012**, *68*, o41–o44. [CrossRef]
16. Colombo, V.; Masciocchi, N.; Palmisano, G. Crystal Chemistry of the Antibiotic Doripenem. *J. Pharm. Sci.* **2014**, *103*, 3641–3647. [CrossRef]
17. Kobayashi, K.; Fukuhara, H.; Hata, T.; Sekine, A.; Uekusa, H.; Ohashi, Y. Physicochemical and Crystal Structure Analyses of the Antidiabetic Agent Troglitazone. *Chem. Pharm. Bull.* **2003**, *51*, 807–814. [CrossRef] [PubMed]
18. Fujii, K.; Aoki, M.; Uekusa, H. Solid-State Hydration/Dehydration of Erythromycin A Investigated by ab Initio Powder X-ray Diffraction Analysis: Stoichiometric and Nonstoichiometric Dehydrated Hydrate. *Cryst. Growth Des.* **2013**, *13*, 2060–2066. [CrossRef]

19. Fujii, K.; Uekusa, H.; Itoda, N.; Yonemochi, E.; Terada, K. Mechanism of Dehydration-Hydration Processes of Lisinopril Dihydrate Investigated by ab Initio Powder X-ray Diffraction Analysis. *Cryst. Growth Des.* **2012**, *12*, 6165–6172. [CrossRef]
20. Brown, G.W.; Paes, D.; Bryson, J.; Freeman, A.J. The effectiveness of a single intravenous dose of ondansetron. *Oncology* **1992**, *49*, 273–278. [CrossRef] [PubMed]
21. Collin, S.; Moureau, F.; Quintero, M.G.; Vercauteren, D.P.; Evrard, G.; Durant, F. Stereoelectronic requirements of benzamide 5HT3 antagonists. Comparison with D2 antidopaminergic analogues. *J. Chem. Soc. Perkin Trans.* **1995**, *2*, 77–84. [CrossRef]
22. Mizoguchi, R.; Uekusa, H. Elucidating the Dehydration Mechanism of Ondansetron Hydrochloride Dihydrate with a Crystal Structure. *Cryst. Growth Des.* **2018**, *18*, 6142–6149. [CrossRef]
23. Sheldrick, G.M. SHELXT—Integrated space-group and crystalstructure determination. *Acta Crystallogr. Sect. A* **2015**, *A71*, 3–8. [CrossRef]
24. Neumann, M. X-Cell: A novel indexing algorithm for routine tasks and difficult cases. *J. Appl. Crystallogr.* **2003**, *36*, 356–365. [CrossRef]
25. Sun, H.; Jin, Z.; Yang, C.; Akkermans, R.L.C.; Robertson, S.H.; Spenley, N.A.; Miller, S.; Todd, S.M. COMPASS II: Extended coverage for polymer and drug-like molecule databases. *J. Mol. Model.* **2016**, *22*, 47. [CrossRef] [PubMed]
26. Engel, G.E.; Wilke, S.; Konig, O.; Harris, K.D.M.; Leusen, F.J.J. PowderSolve—A complete package for crystal structure solution from powder diffraction patterns. *J. Appl. Crystallogr.* **1999**, *32*, 1169–1179. [CrossRef]
27. Veldhuizen, D.A.V.; Lamont, G.B. Multiobjective Evolutionary Algorithms: Analyzing the State-of-the-Art. *Evol. Comput.* **2000**, *8*, 125–147. [CrossRef] [PubMed]
28. Osaka, K.; Matsumoto, T.; Miura, K.; Sato, M.; Hirosawa, I.; Watanabe, Y. The Advanced Automation for Powder Diffraction toward Industrial Application. *AIP Conf. Proc.* **2010**, *1234*, 9–12. [CrossRef]
29. Jia, Y.Q. Crystal radii and effective ionic radii of the rare earth ions. *J. Solid State Chem.* **1991**, *95*, 184–187. [CrossRef]
30. Chakraborty, S.; Dutta, R.; Arunachalam, M.; Ghosh, P. Encapsulation of [X2(H2O)4]2− (X = F/Cl) clusters by pyridyl terminated tripodal amide receptor in aqueous medium: Single crystal X-ray structural evidence. *Dalton Trans.* **2014**, *43*, 2061–2068. [CrossRef]
31. Hoque, M.N.; Das, G. Overview of the strategic approaches for the solid-state recognition of hydrated anions. *CrystEngComm* **2017**, *19*, 1343–1360. [CrossRef]

© 2019 by the authors. Licensee MDPI, Basel, Switzerland. This article is an open access article distributed under the terms and conditions of the Creative Commons Attribution (CC BY) license (http://creativecommons.org/licenses/by/4.0/).

*Article*

# Solvent-Mediated Polymorphic Transformation of Famoxadone from Form II to Form I in Several Mixed Solvent Systems

### Dan Du, Guo-Bin Ren, Ming-Hui Qi *, Zhong Li and Xiao-Yong Xu *

Shanghai Key Laboratory of Chemical Biology, Laboratory of Pharmaceutical Crystal Engineering & Technology, School of Pharmacy, East China University of Science and Technology, No. 130, MeiLong Road, XuHui District, Shanghai 200237, China; dudanchem@yeah.net (D.D.); rgb@ecust.edu.cn (G.-B.R.); lizhong@ecust.edu.cn (Z.L.)
* Correspondence: mhqi@ecust.edu.cn (M.-H.Q.); xyxu@ecust.edu.cn (X.-Y.X.)

Received: 2 March 2019; Accepted: 13 March 2019; Published: 20 March 2019

**Abstract:** This paper discloses six polymorphs of famoxadone obtained from polymorph screening, which were characterized by XRPD, DSC, and SEM. A study of solvent-mediated polymorphic transformation (SMPT) of famoxadone from the metastable Form II to the stable Form I in several mixed solvent systems at the temperature of 30 °C was also conducted. The transformation process was monitored by Process Analytical Technologies. It was confirmed that the Form II to Form I polymorphic transformation is controlled by the Form I growth process. The transformation rate constants depended linearly on the solubility difference value between Form I and Form II. Furthermore, the hydrogen-bond-donation/acceptance ability and dipolar polarizability also had an effect on the rate of solvent-mediated polymorphic transformation.

**Keywords:** famoxadone; solvent-mediated polymorphic transformation; hydrogen-bond-acceptance ability

---

## 1. Introduction

Crystallization is the widely-used method for isolation pf a solid-state chemical compound from solution at the manufacturing scale, and it is governed and affected by thermodynamic and kinetic factors [1–6]. Polymorphs crystallization often follows Ostwald's rule of stages [7], which postulates that system polymorphic crystallization is carried out from the supersaturated state stage to the equilibrium state in stages. Thus, the metastable form is crystallized first, and then, this system undergoes every possible polymorphic structure before the thermodynamically-stable polymorph crystallizes from the solution [8]. Solvent-mediated polymorphic transformation (SMPT) is a process where the metastable polymorph interacts with the solvent and subsequently transforms to a more stable polymorph by dissolution and crystallization. SMPT is interpreted as a three-step process: dissolution of the metastable polymorph, nucleation of the stable polymorph, and growth of the stable polymorph [9]. The driving force of this process is the difference value between the solubility of the two corresponding polymorphs [10], which consequently determines the supersaturation level during the crystallization process of the thermodynamically-stable form [11]. SMPT has been extensively studied over the years. It is useful for producing the thermodynamically-stable polymorph [12]. On the other hand, in general, if the metastable polymorph is desired as that of a pharmaceutically-active compound, SMPT is also helpful because SMPT is usually a metastable polymorph transform to a stable polymorph, so we need to study the SMPT influencing factor to suppress the SMPT process and obtain the expected metastable polymorph [13]. However, there are no comprehensive studies on the effect and mechanism of solvent in SMPT [14].

Famoxadone (CAS No. 131807-57-3), 3-anilino-5-methyl-5-(4-phenoxyphenyl)-1,3-oxazolidine-2, 4-dione (Figure 1), is a high-efficiency wide-spectrum pesticide produced by DuPont (USA) [15]. Famoxadone is a member of a new class of oxazolidinone fungicides that demonstrate excellent control of plant pathogens in the Ascomycete, Basidiomycete, and Oomycete classes that infect grapes, cereals, tomatoes, potatoes, and other crops [15,16]. There is no literature reporting the polymorphs of famoxadone.

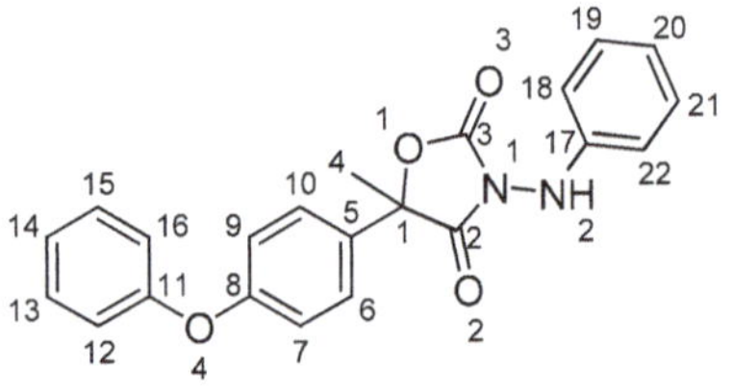

**Figure 1.** Molecular structure of famoxadone.

In this work, six polymorphs were obtained by polymorph screening and defined as Form I–Form VI. The related research showed that Form I is the most thermodynamically-stable form among them. Besides, the SMPT of the famoxadone metastable Form II to the thermodynamically-stable Form I in several mixed solvent systems has been studied, corresponding to mixed solvent systems including nitromethane/toluene, nitromethane/isopropylbenzene, acetone/$m$-xylene, acetone/toluene, acetone/$o$-xylene, acetone/$p$-xylene, and acetone/mesitylene.

## 2. Experimental Section

**Materials.** Form I was obtained by synthesis [16] in the laboratory (purity > 95%, HPLC) and used without further purification. A single crystal of famoxadone Form I was crystallized from methyl tert-butyl ether and $n$-heptane. Acetone, nitromethane, toluene, isopropyl-benzene, $o$-xylene, $m$-xylene, $p$-xylene, and mesitylene of analytical reagent grade (purity > 99%) were purchased from Sinopharm Chemical Reagent Co. Ltd., Shanghai, China. Deionized water was prepared by Ming-che D 24 UV, China.

Form II was prepared by dissolving 100 mg of famoxadone in 1 mL of chloroform followed by adding 10 mL of $n$-heptane. A large amount of white solid precipitated, then the solid powders were filtered and dried over 24 h under vacuum at room temperature.

Form III was prepared by evaporation of the saturated solution of famoxadone in chlorobenzene during two weeks at room temperature. Then, the solid powders were filtered and dried over 24 h under vacuum at room temperature.

Form IV was prepared by dissolving 100 mg of famoxadone in 1 mL $N,N$-dimethylacetamide followed by adding 10 mL of formic acid. A large amount of white solid precipitated immediately, then the solid powders were filtered and dried over 24 h under vacuum at room temperature. Another method for preparing Form IV was from dissolving 100 mg famoxadone in 2 mL of 2-methyltetrahydrofuran followed by adding 20 mL of $n$-butyl ether. A large amount of white solid precipitated after resting for a while, then the solid powders were filtered and dried over 24 h under vacuum at room temperature.

Form V was prepared by dissolving 250 mg of famoxadone in 3 mL of 2-butanone, then adding this solution to 10 mL of 2,2,2-trifluoroethanol at a controlled temperature of 4 °C. A large amount of white solid precipitated immediately, then the solid powders were filtered and dried over 24 h under vacuum at room temperature.

Form VI was prepared by dissolving 100 mg famoxadone in 1 mL of tetrahydrofuran, followed by adding 10 mL of trifluorotoluene. A large amount of white solid precipitated, then the solid powders were filtered and dried over 24 h under vacuum at room temperature.

**Apparatus and Instruments.** *X-ray Technology.* XRPD patterns were obtained using a Rigaku Ultima IV X-ray diffractometer (Cu-Kα radiation) [17]. The voltage and current of the generator were set to 40 kV and 40 mA, respectively. Data over the 2θ angular range of 5°–45° were collected at a scan rate of 20°/min at ambient temperature. The data were imaged and integrated with RINT Rapid, and the peaks were analyzed with Jade 6.0 from Rigaku.

XRPD calibration was performed using the mixtures containing 5%, 10%, 20%, 50%, 80%, 90%, and 95% Form I in Form II (~0.3 g total sample weight). The mixtures were weighed by using an analytical balance, and the samples were heterogeneous solid-state mixed. Quantitative solid phase analyses were performed using fundamental parameter-based Rietveld software BGMN [18]. Structure data necessary for quantitative analysis were acquired from the experimental data of famoxadone Form I and Form II. XRPD patterns of calibration samples and experiment samples collected were recorded and analyzed identically.

**SXRD.** Single-crystal X-ray diffraction for famoxadone Form I were collected on a Bruker SMART-APEX DUO diffractometer (USA) equipped with a graphite monochromator and Mo-Kα fine-focus sealed tube ($\lambda = 0.71073$ Å). Data processing was performed using Bruker SAINT Software. X-ray diffraction intensities were corrected for absorption using SADABS [19], and the structure was refined using SHELXL-97 [20]. All non-hydrogen atoms were refined anisotropically [21]. Hydrogen atoms on heteroatoms were located from different electron density maps, and all C-H hydrogens were fixed geometrically. Hydrogen bond geometries were determined by Platon software. Crystal structures may be accessed at www.ccdc.cam.ac.uk/data_request/cif (CCDC No. 1560380).

*Thermal Analysis.* Differential scanning calorimetry (DSC, Q2000, TA Instruments, New Castle, DE, USA) was used to measure the fusion enthalpies and melting points of the polymorphs. Samples weighing 3–5 mg were placed in crimped and sealed aluminum sample pans and heated from 30 °C–200 °C at a rate of 10 °C/min. The instrument was calibrated against with the melting characteristics of indium by a standard procedure. Each sample was analyzed in triplicate with RSD < 2%.

*Morphology Observation.* The shapes and morphologies of the polymorphs of famoxadone were examined on a Carl Zeiss model Merlin Compact 6027 FESEM (Oberkochen, Germany) with a beam voltage of 3 kV [22]. The FESEM imaging of sample was spread on a carbon-coated copper grid in order to enhance the conductivity.

*Process Analytical Techniques.* FBRM® (Focused Beam Reflectance Measurement) was used to monitor the chord length of particle change. The FBRM® probe (Particle Track G400, Mettler Toledo, Switzerland) with a chord length measurement range of 1–1000 μm and was set as 10-s sampling intervals [23]. The ReactIR™ system was used to calibrate standard solutions and utilized to record the change of concentration of the solution during SMPT processes. The ReactIR™ probe (ReactIR™ 15, Mettler Toledo) was used with 15-s sampling intervals [24]. All transformation experiments were performed in a 100-mL Easy-Max vessel (Mettler Toledo, Greifensee, Switzerland) in conjunction with iControl Easy-max software [25]. The accuracy of temperature control on this system was 0.01 °C.

**Solubility Measurements.** An excess amount of thermodynamically-stable Form I was added to 15 mL of solvent and then left to slurry overnight at 30 ± 1 °C. The saturated solution was filtered through a 0.22-μm syringe filter, and the clear solution was transferred to a preweighed glass vial. The solution was left to evaporate at room temperature and weighed, and the solubility was calculated [26]. To determine the solubility of Form II, a certain amount of solvent was introduced to the vessel previously, and the solution was stirred continuously at 200 rpm under 30 °C. Then, a small amount of Form II was added gradually, and a small amount was continuously added after Form II completely dissolved, the mass of each addition being about 2 mg. If Form II was not completely dissolved within 30 min after the last addition of it, the solution was regarded as a saturated one. The total addition was recorded, and the range of the solubility could be determined. Two parallel experiments were performed for each measurement.

**Slurry Experiments.** Slurry experiments were designed to investigate the influence of solvent conditions in the solvent-mediated polymorphic transformation (SMPT) process.

SMPT was investigated in nitromethane/toluene (1:1), nitromethane/isopropyl benzene (1:1), acetone/*m*-xylene (1:1), acetone/toluene (1:1), acetone/*o*-xylene (1:1), acetone/*p*-xylene (1:1), and acetone/mesitylene (1:1). For each experiment, about 60 mL of saturated solution of Form I were added to the vessel under 30 °C, and then, preweighed Form II was added to this solution. Agitation speed was set at 200 rpm, and the system temperature was controlled by Easy-max. In addition, it was assumed that the insertion of PAT analyzers had no influence on the nature of the transformation mechanism.

The solid polymorphic composition (~10 mg) was determined by using a disposable plastic dropper from the suspension, and this collected solid phase was quickly filtered through 2–3-μm filter paper with a Büchner flask under reduced pressure. The quantity of Form I in the sample was monitored and quantified via ex situ powder X-ray diffraction (XRPD) analysis (Figure S1 in the supplementary materials).

## 3. Results and Discussion

### 3.1. The Characterization of Starting Materials

**The characterization of starting materials.** The six polymorphs were readily distinguished by their unique XRPD patterns (Figure 2) and can be identified by the diagnostic peaks for Form I at $2\theta$ = 6°, 7.9°, 9.6°, 10.2°, 12°, 14.8°, 17.3°, 18.5°, 19.2°, 19.8°, and 20.5°; for Form II at $2\theta$ = 10°, 16.7°, 18.1°, 20.2°, 20.8°, 21.2°, 22.2°, 24.3°, and 28.5°; for Form III at $2\theta$ = 4.9°, 9.2°, 13.4°, 15.9°, 18°, 19.6°, 20.5°, 21.2°, and 22.3°; for Form IV at $2\theta$ = 7°, 9.5°, 10.4°, 18.5°, 19.1°, 19.7°, 21.2°, 23.1°, and 28.8°; for Form V at $2\theta$ = 5.2°, 10.1°, 11.5°, 14.8°, 15.9°, 18.6°, 20°, 20.9°, and 23.5°; for Form VI at $2\theta$ = 6.3°, 8.7°, 9.1°, 12.5°, 17.2°, 18.1°, 18.6°, 20.9°, and 25.8°. Images of SEM of the polymorphs are shown in Figure 3. Detailed data of the peaks in PXRD patterns are listed in Tables S1–S6.

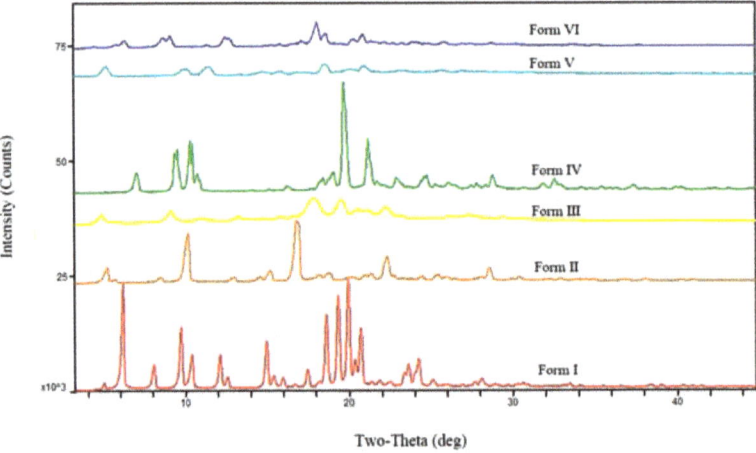

**Figure 2.** XRPD patterns of famoxadone from Form I to Form VI polymorphs.

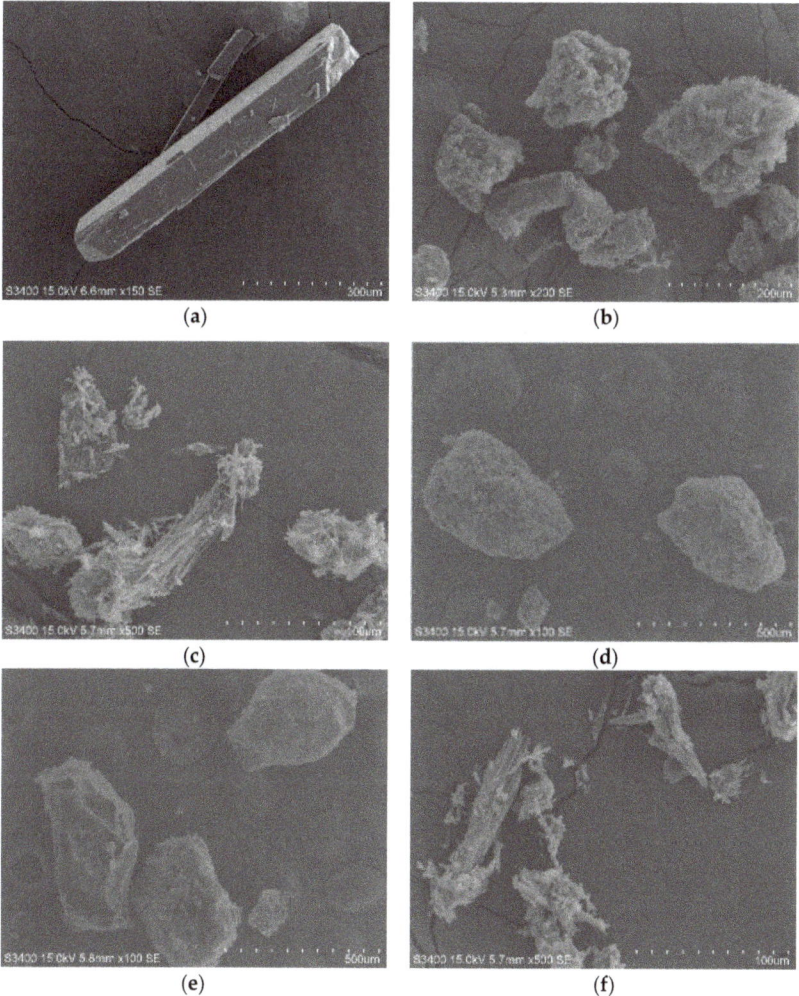

**Figure 3.** SEM images of famoxadone Form I (**a**), Form II (**b**), Form III (**c**), Form IV (**d**), Form V (**e**), and Form VI (**f**).

The crystal structure of Form I was determined by the orthorhombic $P2_12_12_1$ space group ($Z = 4$, $Z' = 1$) (Figure 4). Crystallographic data of Form I are summarized in Table 1. There were hydrogen-bonding interactions between O(3) of the carbonyl-group and N(2) of the imino group in adjacent Form I molecules. The hydrogen bonding distance of N(2)–H···O(3) was 2.512 Å. There was one literature report about the crystal structure of (S)-famoxadone [16]; the crystallographic data in the literature and the data obtained in this paper were compared, and it was found that the crystallographic data parameters of these two forms were approximately the same.

Table 1. Crystallographic information of famoxadone Form I.

| Polymorph | Form I |
|---|---|
| Crystal system | Orthorhombic |
| Space group | $P2_12_12_1$ |
| a (Å) | 5.8880 (5) |
| b (Å) | 16.9635 (16) |
| c (Å) | 18.4218 (17) |
| α (°) | 90 |
| β (°) | 90 |
| γ (°) | 90 |
| Volume (Å$^3$) | 1840.0(3) |
| Z | 4 |
| CCDC No. | 1560380 |
| Temperature | 130 K |
| Crystal size | $0.22 \times 0.08 \times 0.05$ mm$^3$ |
| Theta range for data collection | 1.632 to 30,565° |
| Reflections collected | 18,744 |
| Absorption correction | Semi-empirical from equivalents |
| Final R indices [$I>2\sigma$ (I)] | $R_1 = 0.0395, \omega R_2 = 0.0932$ |

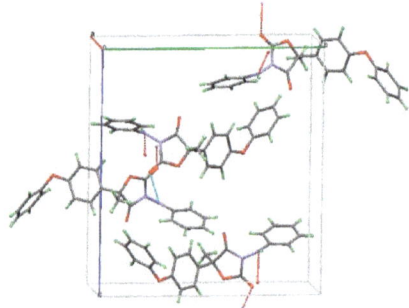

Figure 4. Famoxadone racemic Form I crystal packing projection graphs.

Polymorphs of famoxadone were also detected by DSC and are shown in Figure 5. The melting data are listed in Table 2. According to the heat of fusion rule [27], it was demonstrated that Form I is the most thermodynamically-stable polymorph. Based on the relevant thermodynamic data and phenomenon, Form I and Form II are in monotropic system [28]; thus, the transformation from Form II to Form I should proceed spontaneously. The stability experiments of the six famoxadone polymorphs are shown in Figures S2–S4. The experimental results showed that six polymorphs were stable under high temperature, high humidity, and strong light conditions for ten days and without polymorph change.

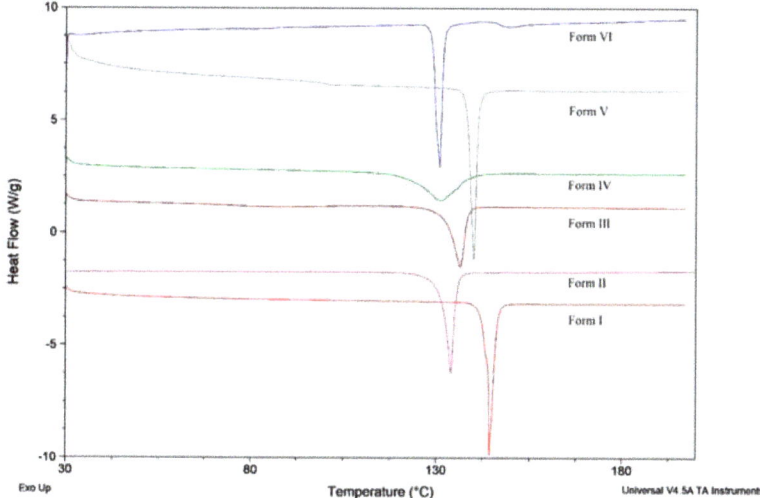

**Figure 5.** DSC curves of famoxadone from Form I to Form VI.

**Table 2.** Melting data of famoxadone polymorphs.

| Polymorph | $T_{onset}$ (°C) | $T_{peak}$ (°C) | $H_f$ (kJ/mol) |
|---|---|---|---|
| Form I   | 141.86 | 144.07 | 93.43 |
| Form II  | 129.78 | 133.95 | 91.27 |
| Form III | 130.63 | 136.51 | 71.89 |
| Form IV  | 119.99 | 131.2  | 80.23 |
| Form V   | 137.12 | 140.22 | 88.02 |
| Form VI  | 128.72 | 130.87 | 86.10 |

### 3.2. Solvent-Mediated Polymorphic Transformation Experiments

In the actual research process, it was found that Form I and Form II were easy to prepare and had good commercial application value. Therefore, a series of experiments was carried out to investigate the transformation process of Form II to Form I, and the basic experimental parameters are listed in Table 3. The mass of Form II added in each experiment was determined by initial supersaturation equivalent, which refers to the same ratio of the actual concentration of the solute that exceeds the saturated concentration over the saturated concentration. The formula was defined as $\sigma = (C - C_0)/C_0$, where $\sigma$ is supersaturation, $C$ is the actual concentration and $C_0$ is the saturated concentration.

**Table 3.** Experimental conditions and parameters.

| No. | Mixed Solvent System | $T$ (°C) | Approximate Form II Added at $t = 0$ min (g) | Approximate $t_{trans}$ (min) |
|---|---|---|---|---|
| #1 | Nitromethane/toluene (1:1) | 30 | 2.06 | 1 |
| #2 | Nitromethane/isopropylbenzene (1:1) | 30 | 1.51 | 2 |
| #3 | Acetone/$m$-xylene (1:1) | 30 | 0.49 | 3 |
| #4 | Acetone/toluene (1:1) | 30 | 5.57 | 5 |
| #5 | Acetone/$o$-xylene (1:1) | 30 | 5.72 | 15 |
| #6 | Acetone/$p$-xylene (1:1) | 30 | 5.31 | 19 |
| #7 | Acetone/mesitylene (1:1) | 30 | 1.23 | 26 |

$t_{trans}$ means the required time from the start of the addition of Form II to the completion of the transformation to Form I.

Experiment #4 was chosen as an example to study the transformation. The relevant variation curves of particle counts are shown in Figure 6. The solution concentration of famoxadone in the slurry is shown in Figure 7. The proper amount of solids was sampled from the slurry at different times and examined by XRPD, and the relevant patterns are shown in Figure 8. Other experiments' relevant variation curves of particle counts are shown in Figures S5–S10.

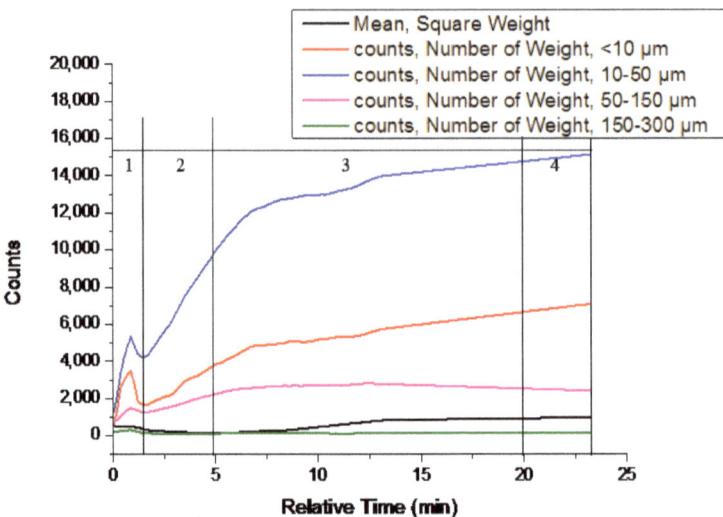

Figure 6. Variation curve of particle counts and mean of chord length in Experiment #4.

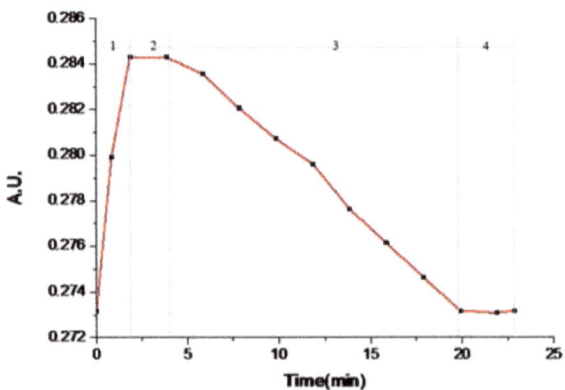

Figure 7. Solution concentration change of famoxadone in Experiment #4.

At the beginning of the experiment, there was a process of adding Form II as Stage (1) in Figure 7 showed: the solution concentration of famoxadone increased rapidly until the dissolution equilibrium of Form II was reached, which was recorded by ReactIR$^{TM}$. As the Stage (1) curve of Figure 6 shows, the particle counts of suspensions increased firstly and then decreased because of the adding of Form II and subsequently the dissolution of it. There was no polymorph change in Stage (1), which can be seen from the XRPD patterns.

In Stage (2), the nucleation and growth of Form I occurred. As Figure 6 shows, from $t = 1.4$ min, the 0–150-μm counts increased, while the 150–300-μm counts decreased. The mean chord length decrease meant the appearance of some small crystal nuclei. The characteristic peak of Form I ($2\theta = 19.78 \pm 0.2°$) was first observed in the XRPD pattern of the separated sample, which meant the suspensions solid phase was the mixture of Form I and Form II, as the XRPD pattern shown in

Figure 8. The solution concentration remained at a platform, and the continual dissolution of metastable Form II would supply the driving force of the nucleation and growth of stable Form I. Cardew and Dawey [29] proposed a theory by means of the solution concentration platform value to deduce the rate-determining step. If the solution concentration platform value is near the metastable form's solubility, the nucleation and growth of the stable form would determine the rate of SMPT. Conversely, the dissolution process of the metastable form would be the rate-determining step when the platform value is near to the solubility of the stable form. In this experiment, the solution concentration platform value was approximately equal to the solubilities of metastable Form II through the conversion between the ReactIR$^{TM}$ intensity and the corresponding solution concentration, and the induction times for all experiments were very short. From these, it can be concluded that the growth process of Form I is the rate-determining step.

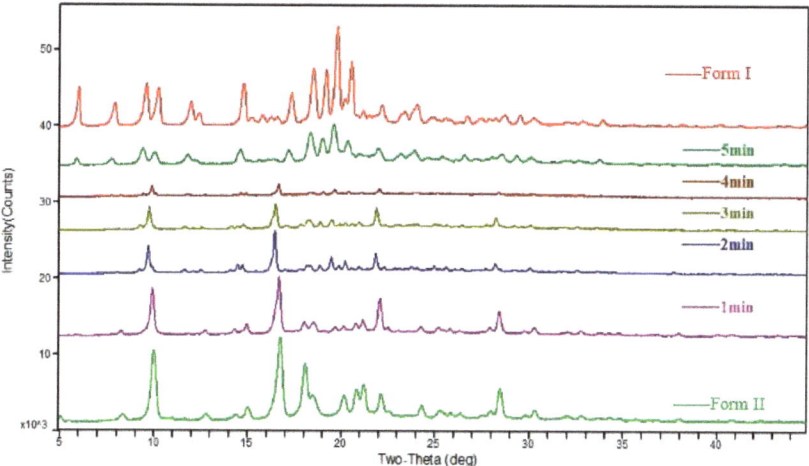

**Figure 8.** Comparison diagram of XRPD during the transformation process in Experiment #4.

After $t = 5$ min, it would be considered as Stage (3). The suspended Form II disappeared, and only Form I existed, as shown in the XRPD patterns in Figure 8. In this period, the mean chord length increased as shown in Figure 6, and this was due to the growth of Form I. As Form II was completely consumed, with the growth of Form I, the solution concentration and supersaturation level began to decrease rapidly, as shown in Figure 7.

Stage (4) is the end of transformation process. In this period, the solution concentration remained invariant, and the mean chord length increased due to the secondary process (aging, aggregation, breakage, etc.) of Form I.

The results of the kinetic experiments performed at the experimental temperature (30 ± 1 °C) represent a series of plots documenting the composition of the solid polymorph during SMPT, as detected by the XRPD method. Figure 8 shows the composition of the solid sampled from the slurry during the SMPT in the acetone/toluene (1:1) system. The reaction rates of each mixed solvent system were compared using this approach. This figure clearly demonstrates that the polymorphic transformation models were similar in all of the solvents, and the only factor that changed was the polymorphic transformation rate.

Considering that the limiting step in nucleation and growth is the transportation of the famoxadone molecule to the crystallization zone, famoxadone molecules diffused at a constant rate during the course of polymorph transition. Besides, the polymorph transition from Form II to Form I accorded with the classical nucleation theory [13] and because it occurred in the saturated solution of Form I. Taking all this into account, the nucleation and growth rate can be assumed to be constant. In order to compare SMPT rates quantitatively in different mixed solvent systems, experimental data were fitted with an appropriate kinetic model [30,31]. Solid-state kinetic models are theoretical, mathematical descriptions of experimental data and are usually expressed as rate equations [32]. The experimental data were in agreement with those for power solid-sate kinetic models [33] (Figure S11). The best correlation was observed when the experimental data points were fitted to the power model P2:

$$\alpha = (kt)^2 \qquad (1)$$

where $k$ is the nucleation and growth rate constant, $\alpha$ is the weight fraction of Form I in the sample, and $t$ is time. The correlations of the experimental data with the theoretical model in each solvent are given in Table S7. Rate constants were calculated based on the least squares method by using MS Excel Solver optimization software [34]. Corresponding nucleation growth rate constants $k$ for each set of experiment calculated are listed in Table 4.

Table 4. The rate constant and solubility data of Form I and Form II in experiments at 30 °C.

| No. | Mixed Solvent System | Rate Constant $k$ (min$^{-1}$) | $S_{Form\ I}$ (g/L) | $S_{Form\ II}$ (g/L) |
|---|---|---|---|---|
| #1 | nitromethane/toluene (1:1) | 5.98 | 0.15555 | 0.15609 |
| #2 | nitromethane/isopropylbenzene (1:1) | 3.26 | 0.29251 | 0.29278 |
| #3 | acetone/m-xylene (1:1) | 2.78 | 0.26269 | 0.26293 |
| #4 | acetone/toluene (1:1) | 1.73 | 0.19559 | 0.19577 |
| #5 | acetone/o-xylene (1:1) | 0.55 | 0.28751 | 0.28755 |
| #6 | acetone/p-xylene (1:1) | 0.46 | 0.26279 | 0.26281 |
| #7 | acetone/mesitylene (1:1) | 0.31 | 0.49958 | 0.49959 |

The solubility data of Form I and Form II are also listed in Table 4. The correlation of SMPT rate constant and the difference value between Form I and Form II solubilities were analyzed (Figure 9). This trend in SMPT from Form II to Form I is described by the equation:

$$k = 10790.8819\ \Delta S + 0.1532 \qquad (2)$$

where $\Delta S$ is the solubility difference value between Form I and Form II and $k$ is the rate constant for SMPT. The correlation coefficient was $R^2 = 0.991$. This means that there was a good correlation between the SMPT driving force solubility difference value $\Delta S$ and the nucleation growth rate constant $k$.

According to the classical nucleation theory, the metastable polymorph has higher nucleation rate in a solvent with higher solubility, and thus, the polymorphic transformation rate is faster [13]. However, the polymorphic transformation from Form II to Form I in this experiment obtained non-conforming results, indicating that the solute–solvent interaction also may have had an important effect on crystallizing kinetics.

The solute–solvent interaction may influence the nucleation and crystal growth rate in two ways. Firstly, the solvent interacted with dissolved molecules, which is called solvation. During the nucleation and crystal growth step, the solvated solute must precede the desolvation process and then integrate into the stable polymorph crystal lattice. Secondly, the solvent molecules are adsorbed on the surface of nuclei or a growing crystal cluster. The incoming solute molecules must replace the solvent molecules to become a part of stable polymorph crystal lattice [35]. For these two reasons, the stronger the solute–solvent interactions, the greater the retardation of nucleation and crystal growth and the slower the polymorphic transformation rate in the solution [36,37]. In this case, the famoxadone nucleation and crystal growth process was affected by the nature of the mixed solvent systems.

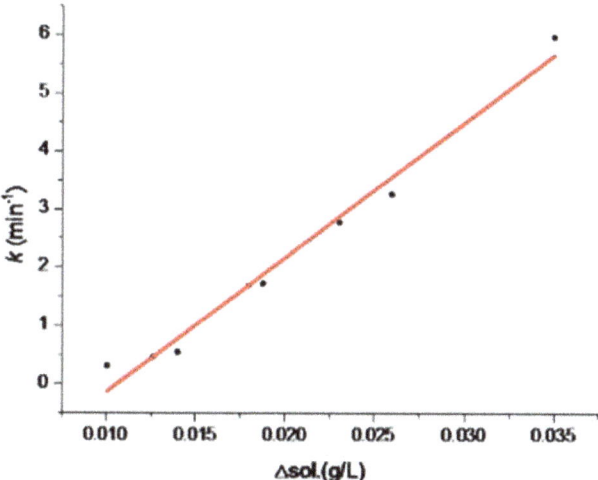

**Figure 9.** Correlation of SMPT rate constants ($k$) and the difference value between the solubilities of Form I and Form II ($\Delta S$) in mixed solvent system.

The solute–solvent interactions are mainly composed of van der Waals forces and hydrogen bonding [13]. The strength of van der Waals interactions between solute and solvent is determined by the dipolar polarizability, $\pi^*$ [38]. The strength of solute–solvent hydrogen-bonding interactions is determined by the hydrogen-bond-donation (HBD) ability $\alpha$ and the hydrogen-bond-acceptance (HBA) ability $\beta$ [39]. The values of $\pi^*$ and $\beta$ of solvents used in the experiments are listed in Table 5. Considering the hydrogen bond combination mode, the imino group in famoxadone has the biggest advantage to form hydrogen-bond as a hydrogen-bond-donor. Therefore, the strongest hydrogen-bond-receptor ability mixed solvent system will have the strongest solute–solvent interactions with famoxadone and retard the polymorph transformation process. In Experiments #1 and #4, the difference of the solvent was nitromethane and acetone. They have similar $\pi^*$, but different $\beta$; acetone has a stronger hydrogen-bond-acceptance ability, so it has strong solute–solvent interactions with famoxadone, leading to a longer transformation time in Experiment #4. For Experiments #3–#7, the acetone was identical, and we only needed to compare another solvent. Toluene, xylene, and mesitylene all have similar $\beta$, so the dominant role is the strength of $\pi^*$. Different solvents have different van der Waals interaction strengths with famoxadone, which will affect the time of polymorph transformation. It is noteworthy that Experiments #3 and #4 had opposite experimental results, and this may be due to the effect of the viscosity ($\mu$) of the solvent, $\mu$ ($m$-xylene) = 0.7500 mPa s (25 °C), $\mu$ (toluene) = 0.5866 mPa s (25 °C). Viscosity may affect solvent-mediated polymorph transformation by affecting the mass transfer process [40,41].

**Table 5.** Values of the hydrogen-bond-acceptance (HBA) ability $\beta$ and polarity/polarizability $\pi^*$ of nitromethane, acetone, toluene, $o$-xylene, $m$-xylene, $p$-xylene, and mesitylene [38,39].

| Solvent | Nitromethane | Acetone | Toluene | $o$-Xylene | $m$-Xylene | $p$-Xylene | Mesitylene |
|---|---|---|---|---|---|---|---|
| $\beta$ | 06 | 43 | 11 | 12 | 12 | 12 | 13 |
| $\pi^*$ | 85 | 71 | 54 | N/A | 47 | 43 | 41 |

## 4. Conclusions

In this paper, six polymorphs of famoxadone were prepared and characterized through XRPD and thermal analysis. In the polymorphic screening process, we only obtained the single crystal structure of Form I, and the other five polymorph single crystals were not obtained by careful cultivation. The single crystal data obtained in this paper were compared with the literature data. The solvent-mediated polymorphic transformation of famoxadone from Form II to Form I in several mixture solvent systems at 30 °C was also studied. The results show that Form I is the most stable form among the polymorphs from the thermodynamic point of view. The SMPT process could be divided into four periods, and was controlled by the growth of the stable Form I. The rate constants for this process were in the range from 0.31 min$^{-1}$–5.98 min$^{-1}$ and depended linearly on the solubility difference value between the two polymorphs.

It is concluded that the acetone/mesitylene mixture solvent system had the strongest interactions with famoxadone molecules compared to the others because of its higher hydrogen-bond-acceptance ability and lower dipolar polarizability, thereby retarding the nucleation and growth of Form I during the transformation process.

**Supplementary Materials:** The following is available at http://www.mdpi.com/2073-4352/9/3/161/s1. Additional data and figures: Crystallographic information, descriptions of some experimental details, tables, and graphs. Accession Codes: CCDC 1560380 contains the supplementary crystallographic data for this paper. These data can be obtained free of charge via www.ccdc.cam.ac.uk/data_request/cif, or by emailing data_request@ccdc.cam.ac.uk, or by contacting The Cambridge Crystallo-graphic Data Centre, 12, Union Road, Cambridge CB2 1EZ, U.K.; fax: +44 1223 336033.

**Author Contributions:** D.D. participated in the design of this study, collected important background information, literature search, performed the statistical analysis, acquisition of data, data analysis and interpretation, and drafted the manuscript. G.-B.R. carried out the study and guided background information; M.-H.Q. participated in the design of this study, data analysis and interpretation, and revised the manuscript. Z.L. carried out the study and guided background information. X.-Y.X. participated in the design of this study, carried out the concepts, and performed manuscript review and revise.

**Funding:** This work was supported by the National Key Research Program of China (No. 2017YFD0200505), the National Natural Science Foundation of China (No. 21706064), and the Fundamental Research Funds for the Central Universities (No. 222201718004, No. 222201814049).

**Conflicts of Interest:** The authors declare no competing financial interest.

## References

1. Maciej, P.; Piotr, C.; Maciej, P.; Dorota, Z.; Miroslaw, K. On the origin of surface imposed anisotropic growth of salicylic and acetylsalicylic acids crystals during droplet evaporation. *J. Mol. Model.* **2015**, *21*, 49.
2. Gagniere, E.; Mangin, D.; Puel, F. Formation of co-crystals: Kinetic and thermodynamic aspects. *J. Cryst. Growth* **2009**, *311*, 2689–2695. [CrossRef]
3. Cysewski, P.; Przybyłek, M.; Miernik, T.; Kobierski, M.; Ziółkowska, D. On the origin of surfaces-dependent growth of benzoic acid crystal inferred through the droplet evaporation method. *Struct. Chem.* **2015**, *26*, 705–712. [CrossRef]
4. Terry, T. Crystallisation of Polymorphs: Thermodynamic Insight into the Role of Solvent. *Org. Process Res. Dev.* **2000**, *4*, 384–390.
5. Bhugra, C.; Pikal, M.J. Role of thermodynamic, molecular, and kinetic factors in crystallization from the amorphous state. *J. Pharm. Sci.* **2008**, *97*, 1329–1349. [CrossRef] [PubMed]
6. Clavaguera, N.; Clavagueramora, M.T.; Crespo, D.; Pradell, T. Thermodynamic and kinetic factors driving primary crystallization. In Proceedings of the Fifth International Workshop on Non-Crystalline Solids, Santiago de Compostela, Spain, 2–5 July 2015; pp. 272–277.
7. Stoica, C.; Tinnemans, P.; Meekes, H.; Vlieg, E.; Hoof, P.J.C.M.V.; Kaspersen, F.M. Epitaxial 2D Nucleation of Metastable Polymorphs: A 2D Version of Ostwald's Rule of Stages. *Cryst. Growth Des.* **2005**, *5*, 975–981. [CrossRef]
8. Chen, C.; Cook, O.; Nicholson, C.E.; Cooper, S.J. Leapfrogging Ostwald's Rule of Stages: Crystallization of Stable γ-Glycine Directly from Microemulsions. *Cryst. Growth Des.* **2011**, *11*, 2228–2237. [CrossRef]
9. Murphy, D.; RodríGuez-Cintrón, F.; Langevin, B.; Kelly, R.C.; RodríGuez-Hornedo, N. Solution-mediated phase transformation of anhydrous to dihydrate carbamazepine and the effect of lattice disorder. *Int. J. Pharm.* **2002**, *246*, 121. [CrossRef]
10. Nguyen, D.L.T.; Kim, K.J. Solvent-Mediated Polymorphic Transformation of α-Taltirelin by Seeded Crystallization. *Chem. Eng. Technol.* **2016**, *39*, 1281–1288. [CrossRef]

11. Liu, Y.; Gao, H.; Xu, H.; Ren, F.; Ren, G. Influence of Temperature, Solvents, and Excipients on Crystal Transformation of Agomelatine. *Org. Process Res. Dev.* **2016**, *20*, 1559–1565. [CrossRef]
12. Yang, L.; Hao, H.; Zhou, L.; Wei, C.; Hou, B.; Xie, C.; Yin, Q. Crystal Structures and Solvent-Mediated Transformation of the Enantiotropic Polymorphs of 2,3,5-Trimethyl-1,4-diacetoxybenzene. *Ind. Eng. Chem. Res.* **2013**, *52*, 17667–17675. [CrossRef]
13. Gu, C.H.; Young, V.; Grant, D.J.W. Polymorph screening: Influence of solvents on the rate of solvent-mediated polymorphic transformation. *J. Pharm. Sci.* **2001**, *90*, 1878–1890. [CrossRef] [PubMed]
14. Schöll, J.; Bonalumi, D.; Lars Vicum, A.; Mazzotti, M.; Müller, M. In Situ Monitoring and Modeling of the Solvent-Mediated Polymorphic Transformation of l-Glutamic Acid. *Cryst. Growth Des.* **2006**, *6*, 881–891. [CrossRef]
15. Sternberg, J.A.; Geffken, D.; Adams, J.B.; Pöstages, R.; Sternberg, C.G.; Campbell, C.L.; Moberg, W.K. Famoxadone: The discovery and optimisation of a new agricultural fungicide. *Pest Manag. Sci.* **2001**, *57*, 143–152. [CrossRef]
16. Zheng, Y.-J.; Shapiro, R.; Marshall, W.J.; Jordan, D.B. Synthesis and structural analysis of the active enantiomer of famoxadone, a potent inhibitor of cytochrome $bc_1$. *Bioorg. Med. Chem. Lett.* **2000**, *10*, 1059–1062. [CrossRef]
17. Macaluso, R.T.; Wells, B.; Wangeline, C.; Cochran, K.; Greve, B.K. Powder X-ray diffraction and electron microscopy studies of polycrystalline Au2PrIn. *Polyhedron* **2016**, *114*, 313–316. [CrossRef]
18. Oishi, R.; Yonemura, M.; Nishimaki, Y.; Torii, S.; Hoshikawa, A.; Ishigaki, T. Rietveld analysis software for j-parc. *Nucl. Instrum. Methods Phys. Res.* **2009**, *600*, 94–96. [CrossRef]
19. Fedushkin, I.L.; Nevodchikov, V.I.; Bochkarev, M.N.; Dechert, S.; Schumann, H. Reduction of 2,5-di-tert-butylcyclopentadienone and pyridine with thulium diiodide. Structures of the complexes TmI2(THF)2[η5-But2C5H2O]TmI2(THF)3 and [TmI2(C5H5N)4]2(μ2-N2C10H10). *Russ. Chem. Bull.* **2003**, *52*, 154–159. [CrossRef]
20. Beghidja, C.; Rogez, G.; Kortus, J.; Wesolek, M.; Welter, R. Very strong ferromagnetic interaction in a new binuclear mu-methoxo-bridged Mn(III) complex: Synthesis, crystal structure, magnetic properties, and DFT calculations. *J. Am. Chem. Soc.* **2006**, *128*, 3140–3141. [CrossRef]
21. Powell, D.R. Review of X-Ray Crystallography. In *X-Ray Crystallography*; Girolami, G.S., Ed.; University Science Books: Mill Valley, CA, USA, 2015; 300p, ISBN 9781891389771. (hardcover).
22. Dieterich, M.; Kutchko, B.; Goodman, A. Characterization of Marcellus Shale and Huntersville Chert before and after exposure to hydraulic fracturing fluid via feature relocation using field-emission scanning electron microscopy. *Fuel* **2016**, *182*, 227–235. [CrossRef]
23. Xu, X.; Abhay, G.; Sayeed, V.A.; Khan, M.A. Process analytical technology to understand the disintegration behavior of alendronate sodium tablets. *J. Pharm. Sci.* **2013**, *102*, 1513–1523. [CrossRef]
24. Kennemur, J.G.; Desousa, J.D.; Martin, J.D.; Novak, B.M. Reassessing the Regioregularity of *N*-(1-Naphthyl)-*N'*-(n-octadecyl)polycarbodiimide Using Solution Infrared Spectroscopy. *Macromolecules* **2015**, *44*, 5064–5067. [CrossRef]
25. Costa, I.C.R.; Leite, S.G.F.; Leal, I.C.R.; Miranda, L.S.M.; Souza, R.O.M.A.D. Thermal effect on the microwave assisted biodiesel synthesis catalyzed by lipases. *J. Braz. Chem. Soc.* **2011**, *22*, 1993–1998. [CrossRef]
26. Hamilton, D.L.; Oxtoby, S. Solubility of Water in Albite-Melt Determined by the Weight-Loss Method. *J. Geol.* **1986**, *94*, 626–630. [CrossRef]
27. Chen, J.; Sarma, B.; Evans, J.M.B.; Myerson, A.S. Pharmaceutical Crystallization. *Cryst. Growth Des.* **2011**, *11*, 887–895. [CrossRef]
28. Näther, C.; Jess, I.; Seyfarth, L.; Bärwinkel, K.; Senker, J. Trimorphism of Betamethasone Valerate: Preparation, Crystal Structures, and Thermodynamic Relations. *Cryst. Growth Des.* **2015**, *15*, 366–373. [CrossRef]
29. Cardew, P.T.; Davey, R.J. The Kinetics of Solvent-Mediated Phase Transformations. *Proc. R. Soc. Lond.* **1985**, *398*, 415–428. [CrossRef]
30. Skrdla, P.J.; Robertson, R.T. Dispersive kinetic models for isothermal solid-state conversions and their application to the thermal decomposition of oxacillin. *Thermochim. Acta* **2007**, *453*, 14–20. [CrossRef]
31. Liu, W.; Dang, L.; Wei, H. Thermal, phase transition, and thermal kinetics studies of carbamazepine. *J. Therm. Anal. Calorim.* **2013**, *111*, 1999–2004. [CrossRef]
32. Skrdla, P.J. Crystallizations, solid-state phase transformations and dissolution behavior explained by dispersive kinetic models based on a Maxwell-Boltzmann distribution of activation energies: Theory, applications, and practical limitations. *J. Phys. Chem. A* **2009**, *113*, 9329. [CrossRef]

33. Liu, M.; Pourquie, M.J.B.M.; Fan, L.; Halliop, W.; Cobas, V.R.M.; Verkooijen, A.H.M.; Aravind, P.V. The Use of Methane-Containing Syngas in a Solid Oxide Fuel Cell: A Comparison of Kinetic Models and a Performance Evaluation. *Fuel Cells* **2013**, *13*, 428–440. [CrossRef]
34. Takane, Y.; Young, F.W.; Leeuw, J.D. Nonmetric individual differences multidimensional scaling: An alternating least squares method with optimal scaling features. *Psychometrika* **1977**, *42*, 7–67. [CrossRef]
35. Khoshkhoo, S.; Anwar, J. Crystallization of polymorphs: The effect of solvent. *J. Phys. D Appl. Phys.* **1993**, *26*, B90. [CrossRef]
36. Blagden, N.; Davey, R.J.; Lieberman, H.F.; Williams, L.; Payne, R.; Roberts, R.; Rowe, R.; Docherty, R. Crystal chemistry and solvent effects in polymorphic systems Sulfathiazole. *J. Chem. Soc. Faraday Trans.* **1998**, *94*, 1035–1044. [CrossRef]
37. Gidalevitz, D.; Feidenhans'L, R.; Matlis, S.; Smilgies, D.M.; Christensen, M.J.; Leiserowitz, L. Monitoring in situ growth and dissolution of molecular crystals: Towards determination of the growth unit. *Angew. Chem. Int. Ed. Engl.* **1997**, *36*, 955–959. [CrossRef]
38. Marcus, Y. ChemInform Abstract: The Properties of Organic Liquids that are Relevant to Their Use as Solvating Solvents. *Cheminform* **1994**, *25*, 409–416. [CrossRef]
39. A braham, M.H. *ChemInform Abstract: Scales of Solute Hydrogen-Bonding: Their Construction and Application to Physicochemical and Biochemical Processes*; National Academy of Sciences-National Research Council: Washington, DC, USA, 1993; pp. 277–283.
40. Heijnen, J.J.; Riet, K.V. Mass transfer, mixing and heat transfer phenomena in low viscosity bubble column reactors. *Chem. Eng. J.* **1984**, *28*, B21–B42. [CrossRef]
41. Aimar, P.; Field, R. Limiting flux in membrane separations: A model based on the viscosity dependency of the mass transfer coefficient. *Chem. Eng. Sci.* **1992**, *47*, 579–586. [CrossRef]

© 2019 by the authors. Licensee MDPI, Basel, Switzerland. This article is an open access article distributed under the terms and conditions of the Creative Commons Attribution (CC BY) license (http://creativecommons.org/licenses/by/4.0/).

*Article*

# Preparation of Theophylline-Benzoic Acid Cocrystal and On-Line Monitoring of Cocrystallization Process in Solution by Raman Spectroscopy

Yaohui Huang [1], Ling Zhou [1,*], Wenchao Yang [1], Yang Li [1], Yongfan Yang [1], Zaixiang Zhang [1], Chang Wang [1], Xia Zhang [1] and Qiuxiang Yin [1,2,3,*]

1. State Key Laboratory of Chemical Engineering, School of Chemical Engineering and Technology, Tianjin University, Tianjin 300072, China
2. Collaborative Innovation Center of Chemical Science and Engineering (Tianjin), Tianjin 300072, China
3. Key Laboratory Modern Drug Delivery and High Efficiency in Tianjin University, Tianjin 300072, China
* Correspondence: zhouling@tju.edu.cn (L.Z.); qxyin@tju.edu.cn (Q.Y.)

Received: 5 June 2019; Accepted: 25 June 2019; Published: 27 June 2019

**Abstract:** Pure theophylline-benzoic acid cocrystal was prepared via slurry and cooling crystallization in solution to overcome the disadvantages of existing preparation methods. The target cocrystal was characterized by powder X-ray diffraction (PXRD), thermalgravimetric analysis (TGA), differential scanning calorimetry (DSC) and Raman spectroscopy. The slurry and cooling cocrystallization process in solution was monitored via on-line Raman spectroscopy. The results obtained from on-line Raman monitoring can exhibit the transformation process from raw materials (theophylline and benzoic acid) to cocrystal and show the cocrystal formation rate. Comparing each transformation process under different conditions in slurry crystallization, we found that suspension density of raw materials and temperature both have an impact on the theophylline-benzoic acid cocrystal formation rate. It could be concluded that the cocrystal formation rate increased with the increase of suspension density of raw materials. Further under the same suspension density, higher temperature will accelerate theophylline-benzoic acid cocrystal formation. Meanwhile, various data from the cocrystallization process in cooling crystallization, including nucleation time, nucleation temperature and suitable cooling ending point can be gained from results of on-line Raman monitoring.

**Keywords:** cocrystal; solution crystallization; Raman spectroscopy; on-line monitoring; cocrystal formation

## 1. Introduction

Active pharmaceutical ingredients (API) can exist in different solid-state forms, such as polymorphs, salts, solvates, amorphous forms and cocrystals [1]. Cocrystal is a new class of pharmaceutical crystal form, which can improve the physicochemical properties of API effectively without affecting their internal structure [2]. Cocrystals are molecular complexes that contain two or more components together in the same crystal lattice, and their structures are based on hydrogen bonds, π-π stacking, van der Waals forces and other non-covalent bonds [3]. In contrast to salts, cocrystal formation can be envisaged for acidic, basic and neutral APIs to open up new opportunities for the pharmaceutical industry [4]. Cocrystal synthesis has been carried out using a variety of methods, such as slow evaporation [5,6], cooling crystallization [7,8], slurry crystallization [9], and mechanochemical synthesis and so on [10].

Among these synthesis methods, cooling and slurry crystallization are widely used in preparation of pharmaceutical cocrystals because they can achieve larger scale and higher yield of target cocrystals compared with other methods [7–9]. In recent studies, the research of pharmaceutical cocrystal is mainly focused on how to improve physicochemical properties of APIs and enhance their efficacy.

Many articles have been published which are associated with the design, characterization analysis and properties of cocrystals [11–15]. However, the understanding of cocrystals is still far less than that of other solid forms and the understanding of the cocrystallization process in solution remains to be developed. With cooling and slurry crystallization, the cocrystal formation process is convenient to monitor and control by some on-line analysis methods, such as Raman spectroscopy, Fourier transform infrared spectroscopy (FTIR) and focused beam reflectance measurement (FBRM). Raman spectroscopy, as a kind of reliable and effective technique that can detect both solid and liquid phases, is more suitable to monitor cocrystal formation [16–19].

Theophylline (TP) is a drug for asthma therapy and chronic obstructive pulmonary disease treatment [20]. However, in clinical application, the low water solubility limits its bioavailability. For this reason, new molecular compounds containing theophylline need to be developed to expand its clinical application. Therefore, the synthesis of cocrystal is research focused on improving physicochemical properties of theophylline [21–26]. Theophylline has great potential to form cocrystal due to the carbonyl groups and aromatic nitrogen atoms in the structure, which are readily to form hydrogen bonds. Hence, theophylline is a good model molecule to research cocrystal formation and investigate the cocrystallization mechanism. According to the literature, the cocrystals of theophylline have been successfully prepared with several carboxylic acids, such as glutaric acid, oxalic acid, and benzoic acid [10,27,28]. Benzoic acid (BA) occurs naturally in many plants and serves as an intermediate in the biosynthesis of many secondary metabolites [29]. Benzoic acid is a good non-toxic cocrystal coformer because of the carboxyl group in its structure, which can form many kinds of cocrystals with different APIs. The 1:1 theophylline-benzoic acid cocrystal has been reported to be prepared via neat grinding and slow evaporation [27,30]. These two methods are mostly commonly used in cocrystal screening, however are not conducive for robust scaling because of the inherent limitations of the techniques and solution crystallization most widely used to achieve large-scale production of crystals [13,31]. Therefore, solution crystallization need to be explored to prepare pure theophylline-benzoic acid cocrystal.

Our research focuses on preparation and on-line monitoring of the formation process of theophylline-benzoic acid cocrystal. The molecule structure of TP-BA cocrystal at a stoichiometric molar ratio of 1:1 from the Cambridge Crystallographic Data Centre is shown in Figure 1 [27]. In this study, we successfully prepared pure TP-BA cocrystal via solution crystallization (slurry and cooling crystallization) and characterized the solid phase of the theophylline-benzoic acid cocrystal by different analysis methods. In our research, we investigated the TP-BA cocrystal formation process in slurry and cooling crystallization by on-line Raman spectroscopy. Furthermore, the influence of suspension density of raw materials (theophylline and benzoic acid) and temperature on cocrystal formation rate in slurry crystallization was demonstrated. Further, the nucleation temperature and suitable cooling ending point in cooling crystallization can be also achieved by on-line Raman monitoring.

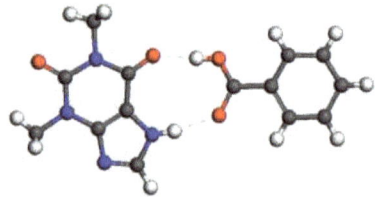

**Figure 1.** The molecules of theophylline-benzoic acid (TP-BA) cocrystal at a stoichiometric molar ratio of 1:1 (grey atoms: C; red atoms: O; blue atoms: N; white atoms: H).

## 2. Materials and Methods

### 2.1. Materials

Theophylline (Form II) was purchased from Aladdin-Reagent Technology Co. Ltd. (Shanghai, China) and benzoic acid was purchased from Tianjin Guangfu Chemical Reagent Co. (Tianjin, China). Deionized water and analytic grade methanol (Tianjin Kewei Chemical Reagent Co., Tianjin, China) were used in this work. All materials were used without further purification. Table 1 shows mass fraction purity and provenance of the materials.

Table 1. Description of materials used in this paper.

| Chemical | Source | Mass Fraction Purity | Purification Method |
|---|---|---|---|
| Theophylline | Aladdin-Reagent Technology Co. Ltd. (Shanghai, China) | >0.990 | GCa |
| Benzoic acid | Tianjin Guangfu Chemical Reagent Co. (Tianjin, China) | >0.990 | GCa |
| Deionized water | Tianjin Kewei Chemical Co. Ltd. (Tianjin, China) | >0.995 | GCa |
| Methanol | Tianjin Kewei Chemical Co. Ltd. (Tianjin, China) | >0.995 | GCa |

GCa: Gas–liquid chromatography.

### 2.2. Preparation of Theophylline-Benzoic Acid Cocrystal

Theophylline-benzoic acid cocrystal was synthesized by slurry and cooling crystallization in methanol/water mixture (V:V = 1:5). In our pre-experiments, the target cocrystal was synthesized in pure acetonitrile and the ratio of API and coformer used for preparation was 1:5 due to the solubility difference between theophylline and benzoic acid, which resulted in benzoic acid being wasted. The mixed solvents (methanol/water mixture) decreased the solubility difference between TP and BA and modified the raw materials ratio to 1:1 to avoid wasting of coformer.

In the slurry crystallization, TP-BA cocrystal solubility in methanol/water mixture solution (V:V = 1:5) was determined by static method with high-performance liquid chromatography (HPLC), which was 0.087 mol/L at 298.15 K. A total of 5.95 mmol TP and 5.9 5mmol BA was added a 36 mL methanol/water mixture to make a suspension of both TP and BA. Due to the total added materials and cocrystal solubility, the suspension density was 0.078 mol/L TP and 0.078 mol/L BA at 298.15 K. The TP and BA suspension was stirred magnetically in a water bath at 298.15 K for at least 5 h to reach equilibrium. Then the cocrystal product was filtered from the suspension and dried at 313.15 K for 12 h. In the cooling crystallization, 3.5 mmol TP and 3.5 mmol BA were dissolved in 36 ml methanol/water mixture (V·V = 1:5). The solution temperature was kept at 313.15 K for 1 h, and then cooled to 278.15 K with a 9 K/h cooling rate. The TP-BA cocrystal can be formed in the cooling process. The solid product of cooling crystallization was isolated over a filter paper (Whatman 2.5 μm grade, Shanghai, China) using vacuum filtration and dried for 12 h at 313.15 K in an oven. The temperature of the water bath was controlled by a thermostat (XODC-2006, Xianou Laboratory Instrument Works Co., Ltd., Nanjing, China), and the system temperature variation for all the measurements was found to be less than ±0.1 K. The cocrystal products gained from slurry and cooling crystallization were analyzed by powder X-ray diffraction (PXRD, Rigaku, Tokyo, Japan) and high-performance liquid chromatography (HPLC, Agilent Technologies, Inc., Carpinteria, CA, USA) to determine solid phase composition. Then TP-BA cocrystal was characterized by differential scanning calorimetry (DSC, Mettler Toledo, Greifensee, Switzerland), thermogravimetric analysis (TGA, Mettler Toledo, Greifensee, Switzerland) and Raman spectroscopy (Kaiser Raman RXN2, Ann Arbor, MI, USA). Raman spectroscopy was used to monitor the cocrystal formation process during slurry and cooling crystallization.

## 2.3. Monitoring Cocrystallization Process of Theophylline and Benzoic Acid in Slurry Crystallization

To explore the factors affecting the TP-BA cocrystal formation process in slurry crystallization, we performed four sets of slurry experiments in a methanol/water mixture (V:V = 1:5). The corresponding temperature, initial concentration and suspension density of raw materials are listed in Table 3. The suspension density was calculated from the total concentration and cocrystal solubility at 298.15 K and 313.15 K. Cocrystal solubility at 298.15 K and 313.15 K was determined by static method with HPLC. The ratio of TP and BA in cocrystal products was analyzed using HPLC to confirm TP-BA cocrystal purity. This cocrystallization process was monitored by Raman spectroscopy for 5 h to reach the equilibrium. The whole spectra was obtained at a spectral range of 200–1800 cm$^{-1}$.

## 2.4. Analytical Methods

A Raman spectrometer (RXN2, Kaiser Optical systems, Inc., Ann Arbor, MI, USA) was used for both off-line measurement of solid samples and on-line monitoring of cocrystal formation process in slurry and cooling crystallization. In the Raman spectroscopy, the spectral resolution and the excitation wavelength of the laser were 5 cm$^{-1}$ and 785 nm, respectively.

The PXRD patterns of the TP–BA cocrystal were obtained using a powder diffractometer (D/MAX 2500, Rigaku, Tokyo, Japan) with a Cu K$\alpha$ radiation (1.54 Å), tube voltage of 40 kV, and current of 100 mA. Data were collected between 2° and 40° in 2$\theta$ with steps of 0.05° and a dwelling time of 1 s per step.

Thermal measurements were performed by TGA/DSC (1/500, Mettler Toledo, Greifensee, Switzerland) protected by nitrogen atmosphere. Experimental conditions were followed as pans of 40 μL volume with a heating rate of 10 K/min and a scan range from 298.15 K to 573.15 K.

The ratio of TP and BA in cocrystal products was analyzed by high-performance liquid chromatography (HPLC). The HPLC is equipped with a UV-vis spectrophotometer detector and uses a C18 column (Extend, 5 μm, 4.6 × 250 mm, Agilent Technologies, Inc., Carpinteria, CA, USA) to separate TP and BA. The mobile phase was composed of 60% methanol and 40% water with 0.1% trifluoroacetic acid. The flow was set at 1 mL/min and the sample injection volume was 20 μL. Absorbance was monitored at 250 nm. Data collection and processing were performed using software from Agilent Technologies, Inc., Carpinteria, CA, USA.

## 3. Results and Discussion

### 3.1. Solid Phases Characterization of Theophylline-benzoic Acid Cocrystal

The powder X-ray diffraction patterns of TP, BA, physical mixture, solid products obtained in slurry and cooling crystallization, and the calculated data of TP-BA cocrystal from single crystal X-ray data [27] are presented in Figure 2a. The PXRD patterns of the cocrystal products obtained in slurry and cooling crystallization are significantly different from those of TP, BA and physical mixture, but the same as the PXRD pattern calculated by single-crystal X-ray data [27]. In detail, the peaks at 6.10° and 17.58°, which are characteristics peaks of TP, are absent in the PXRD patterns of the cocrystal, as is the characteristic peak of BA at 7.98°. Meanwhile, some new peaks appear at 11.40° and 19.46° in the cocrystal product pattern, which are the same as the PXRD pattern of the calculated data of the TP-BA cocrystal. Therefore, the formation of TP-BA cocrystal can be confirmed by the changes in the PXRD patterns. The ratio of TP and BA in cocrystal products is obtained as 1:1 by HPLC.

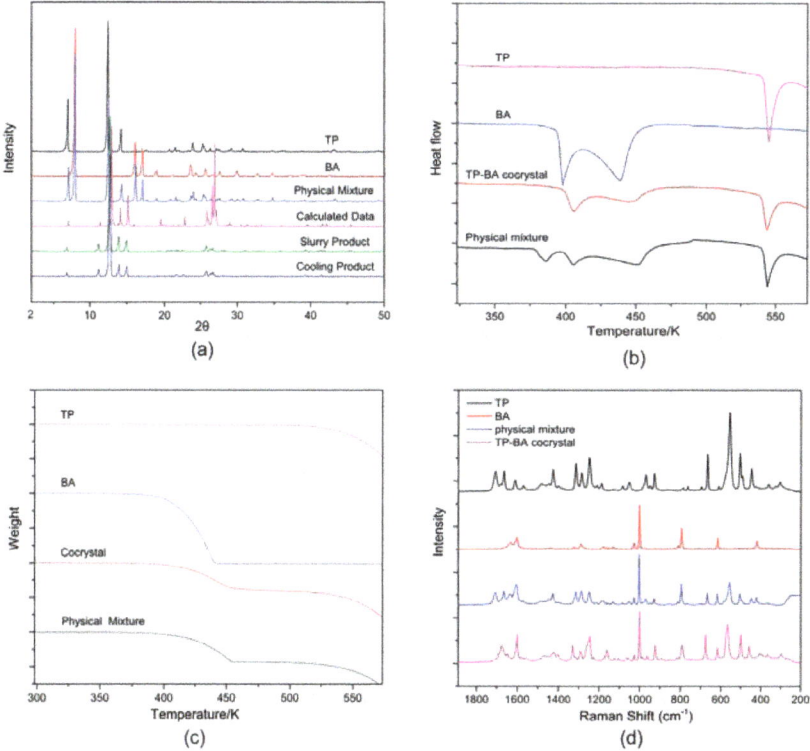

**Figure 2.** The solid characterization spectra of TP, BA, physical mixture and TP-BA cocrystal: (**a**) powder X-ray diffraction (PXRD); (**b**) differential scanning calorimetry (DSC); (**c**) thermogravimetric analysis (TGA); (**d**) Raman spectroscopy.

The DSC and TGA curves of TP, BA, TP-BA cocrystal and physical mixture are shown in Figure 2. The melting points of theophylline and benzoic acid are 545.5 K and 395.1 K, respectively. In the DSC curve of the TP-BA cocrystal, the first endothermic peak at around 411.1 K is the melting point of cocrystal, which is significantly different from API and coformer. The second peak indicates the cocrystal decomposing at 455.3 K. Around this temperature, benzoic acid breaks away from the structure of the TP-BA cocrystal, and theophylline recrystallizes to the solid phase. In the TGA curve of cocrystal, at 455.3 K solid weight begins decreasing and the total weightlessness is about 40%, which is equal to the mass fraction of BA in cocrystal. The last peak in DSC curve of the TP-BA cocrystal at 545.5 K is the melting point of the remaining theophylline. Moreover, the DSC curve of the physical mixture has four endothermic peaks. The peak at 384.5 K indicates that solid theophylline and solid benzoic acid form the TP-BA cocrystal at this temperature and the next peak at 411.1 K is the melting point of the cocrystal, which corresponds with the curve of pure cocrystal. Further, the peaks at 455.3 K and 545.5 K represent the decomposition of BA from the cocrystal and the melting point of the remaining TP, respectively, which are as same as those in the DSC curve of cocrystal.

Figure 2d shows the Raman spectra of the target cocrystal, physical mixture, TP and BA. It is obviously that the spectra of the cocrystal is different from those of TP and BA. For instance, TP and BA have the characteristic peaks at 1688, 1323, 1171 and 918 $cm^{-1}$, while the characteristic peaks of the cocrystal are at 1678, 1331, 1161 and 925 $cm^{-1}$. The difference between the Raman spectra of the cocrystal and raw materials can be used to identify TP-BA cocrystal formation and monitor the cocrystallization process in solution.

## 3.2. Crystal Structure Analysis of TP-BA Cocrystal

The crystallographic data of the TP-BA cocrystal is obtained from Cambridge Crystallographic Data Centre (CCDC) [27] and the crystal structure is shown in Figure 3. Each TP molecule is connected to one BA molecule via two hydrogen bonds to form a dimer. One hydrogen bond is formed between the acidic nitrogen atom on the imidazole ring from the TP molecule and the carboxyl oxygen atom from the BA molecule. The other is formed between the carbonyl group from the TP molecule and the carboxyl group of the BA molecule. The dimers of TP-BA molecules are arranged parallel to the b-c plane, then form stacks along the a-axis. The space group of the TP-BA cocrystal is monoclinic, $P2_1/n$ and has cell parameters $a = 6.98690\ (17)$ Å, $b = 25.10944\ (84)$ Å, $c = 8.60685\ (30)$ Å, $\beta = 108.5597\ (18)$, and $V = 1431.431\ (78)$ Å$^3$. From the TP-BA cocrystal structure, two hydrogen bonds' interaction results in changes in the stretching vibrations of the bonds in TP and BA molecules' structures, which can cause peak shifts in the Raman spectra (Figure 4). For example, the peak at 1688 cm$^{-1}$ is observed in both Raman spectra of TP and the physical mixture, while there is no peak at a similar position in the Raman spectra of BA. It indicates that this peak should be caused by stretching vibrations of the N-C=O bond in TP molecule. The hydrogen bond, N-C=O·H-O, formation, causes this peak to shift from 1688 cm$^{-1}$ to 1678 cm$^{-1}$. Similarly, in the spectra of TP and physical mixture we can see a peak at 918 cm$^{-1}$, which represents N-H on the imidazole ring. After forming a hydrogen bond, N-H·O=C, this peak shifts to 925 cm$^{-1}$. Therefore, during the cocrystallization process, the Raman spectra of the solution would change resulting from hydrogen bonds' interaction between TP and BA molecules, which can be used to monitor the TP-BA cocrystal formation process.

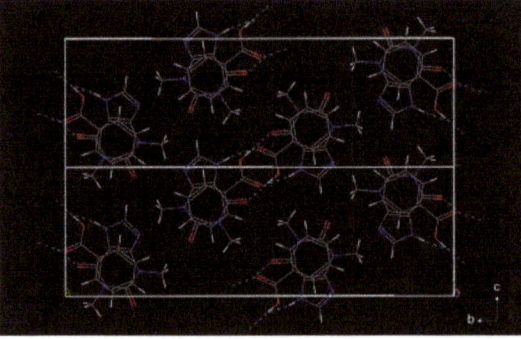

**Figure 3.** The crystal structure of the TP-BA cocrystal (grey color: C; red color: O; blue color: N; white color: H).

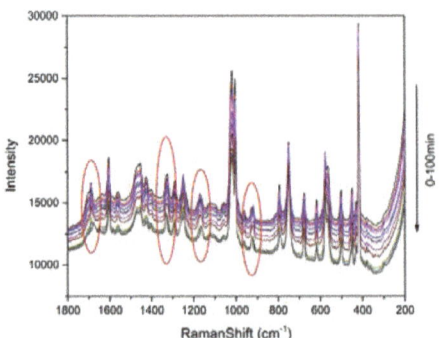

**Figure 4.** The variation of the Raman spectra of TP-BA cocrystal formation in slurry crystallization. The time intervals of the Raman spectra are listed in Table 2.

Table 2. The time intervals of the Raman spectra shown in Figure 4.

| Total Spectra Number | Total Recording Time/min | Time Point of First Spectra/min | Time Interval/min |
|---|---|---|---|
| 20 | 100 | 5 | 5 |

*3.3. On-Line Monitoring of TP-BA Cocrystal Formation Process in Slurry Crystallization*

The Raman spectra at 200–1800 $cm^{-1}$ obtained from slurry crystallization at 298.15 K is shown in Figure 4. In this figure, the shifts of the Raman spectra within operating time can be observed. The main shifts ranges are at (1) 1678–1688 $cm^{-1}$, (2) 1323–1331 $cm^{-1}$, (3) 1161–1171 $cm^{-1}$, and (4) 918–925 $cm^{-1}$, respectively, and are marked with circles in Figure 4, which are corresponding with the off-line Raman spectra of the solid phase of TP, BA and cocrystal. That demonstrates that on-line Raman spectroscopy can be used to monitor the TP-BA cocrystal formation process in slurry crystallization.

Figure 5a presents the Raman spectra of the range of 1640–1720 $cm^{-1}$ obtained from slurry crystallization at 298.15 K. A main peak shift from 1688 $cm^{-1}$ to 1678 $cm^{-1}$ can be observed in the spectra. The solid-state Raman spectra of TP, BA and cocrystal were analyzed in Section 3.1. The characteristic peak at 1688 $cm^{-1}$ can represent raw materials in the suspension, while the peak at the 1678 $cm^{-1}$ could represent the TP-BA cocrystal existing in the slurry. Further, the peak intensity in the Raman spectra is positively related with the concentration of the compounds. The appearance of the characteristic peak, which represents the cocrystal, could demonstrate TP-BA cocrystal nucleation, and the decrease of the characteristic peak intensity of drug and coformer indicates raw material consumption, which is caused by the transformation from raw materials to cocrystal. Therefore, the target cocrystal formation can be identified by monitoring the characteristic peak intensity of the TP-BA cocrystal and raw materials. At the beginning of slurry crystallization only the peak at 1688 $cm^{-1}$ could be seen from the spectra. As the cocrystallization process proceeds, the peak at 1678 $cm^{-1}$ appears and the intensity starts to increase, which proves the TP-BA cocrystal formation. The intensity at 1630 $cm^{-1}$ as the baseline is used to compare the intensity of two peaks, 1678 and 1688 $cm^{-1}$. The difference between the intensity of characteristics peaks and the baseline is named the "relative peak intensity". The changes of the relative peak intensity of 1688 $cm^{-1}$ and 1678 $cm^{-1}$ are shown in Figure 5b. We can see that the relative intensity of 1678 $cm^{-1}$ increases from −250 to 250 over 44 min and then maintains a stable level. Meanwhile, the relative peak intensity of 1688 $cm^{-1}$ decreases from 1500 to 240 during the same period. The results indicate that at this condition the transformation from raw materials to cocrystal finishes at 44 min, and can help us to improve the design of cocrystallization experiments and save more resources and time.

The spectra of 1270–1360 $cm^{-1}$ can be observed in Supplementary Materials Figure S1a. During the cocrystallization process, the intensity of the peak at 1323 $cm^{-1}$ keeps decreasing, while the intensity of the peak at 1331 $cm^{-1}$ gradually increases within the operating time. We choose the intensity of 1306 $cm^{-1}$ as the baseline for the characteristic peaks of 1323 and 1331 $cm^{-1}$. The relative intensity of these two peaks is calculated by the above method and presented in Supplementary Figure S1b. The relative peak intensity of 1323 $cm^{-1}$ decreases from 2000 to 500, while that of the peak at 1331$cm^{-1}$ rises from 1200 to 1750 over a period of about 44 min. Further, the time of transformation from raw materials to cocrystal obtained from the relative intensity change is the same as for the above analysis.

Moreover, two peak shifts can be seen at the range of 1120–1200 $cm^{-1}$ and 890–940 $cm^{-1}$, respectively. One peak at 1171 $cm^{-1}$ shifts to 1161 $cm^{-1}$ (Supplementary Figure S2a) and another moves from 918 to 925 $cm^{-1}$ (Supplementary Figure S3a). When the slurry crystallization begins, we can only see the peak at 1171 and 918 $cm^{-1}$. As the cocrystal formation proceeds, the peaks at 1161 and 925 $cm^{-1}$ appear. The intensities of 1140 and 900 $cm^{-1}$ are used as the baseline. In the Supplementary Figure S2b, the relative intensity of the peak at 1171 $cm^{-1}$ reduces from 1300 to 350 and the relative peak intensity at 1161 $cm^{-1}$ increases from 620 to 1050 over a period of 44 min. Similarly, the relative intensity of the peak at 918 $cm^{-1}$ reduces from 1400 to 300, however the intensity of the peak at 925 $cm^{-1}$ increases from 700 to 1250 at the same time (Supplementary Figure S3b).

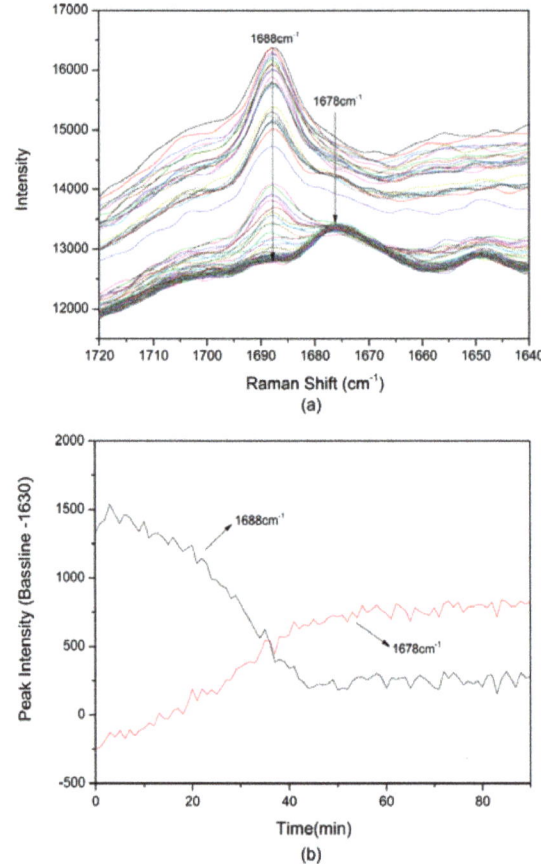

**Figure 5.** (a) The variation of Raman spectra at 1640–1720 cm$^{-1}$; (b) The change of relative peak intensity of 1678/1688 cm$^{-1}$ in TP-BA cocrystal formation in slurry crystallization at 298.15 K. The information about the time interval of the Raman spectra are listed in Table S1.

From the above information obtained from relative intensity change of four couples of characteristics peaks, we know the total time taken for TP-BA cocrystal formation in slurry crystallization under this condition. Therefore, Raman spectroscopy can be used as a reliable and suitable technique to monitor the formation of TP-BA cocrystal and help us to design cocrystallization experiments.

*3.4. Influence of Suspension Density of Raw Materials and Temperature on Theophylline-benzoic Acid Cocrystal Formation in Slurry Crystallization*

To explore the factors affecting TP-BA cocrystal formation process in slurry crystallization, on-line monitoring of the cocrystallization process in solution by Raman spectroscopy was carried out in a methanol/water mixture under different temperatures and suspension densities of raw materials. Besides the above slurry experiment, we added another three sets of experiments at 298.15 K and 313.15 K with different suspension densities of raw materials. The corresponding conditions of the four sets of experiments are shown in Table 3.

Table 3. Temperature, initial concentration, suspension density of raw materials and formation time in TP-BA cocrystal formation process in slurry crystallization.

| | Temperature/K | Initial Concentration of TP/M | Initial Concentration of BA/M | Suspension Density of TP/M | Suspension Density of BA/M | Ratio of TP and BA in Cocrystal [a] | Formation Time/min |
|---|---|---|---|---|---|---|---|
| exp 1 | 298.15 | 0.165 | 0.165 | 0.078 | 0.078 | 1:1 | 44 |
| exp 2 | 298.15 | 0.130 | 0.130 | 0.042 | 0.042 | 1:1 | 57 |
| exp 3 | 313.15 | 0.292 | 0.292 | 0.078 | 0.078 | 1:1 | 22 |
| exp 4 | 313.15 | 0.256 | 0.256 | 0.042 | 0.042 | 1:1 | 43 |

[a] High-performance liquid chromatography (HPLC) chromatograms of product of each set are shown in Supplementary Figure S4.

The peak at 1688 cm$^{-1}$ was chosen as the characterization peak of raw materials in suspension and the peak at 1678 cm$^{-1}$ was selected as the characterization peak of cocrystal. The relative peak intensity is presented in Figure 6 (the baseline is the same as above). In Figure 6, the total time taken for cocrystal formation in each experiment under different conditions can be obtained and the formation time was 44 min, 57 min, 22 min, and 43 min respectively.

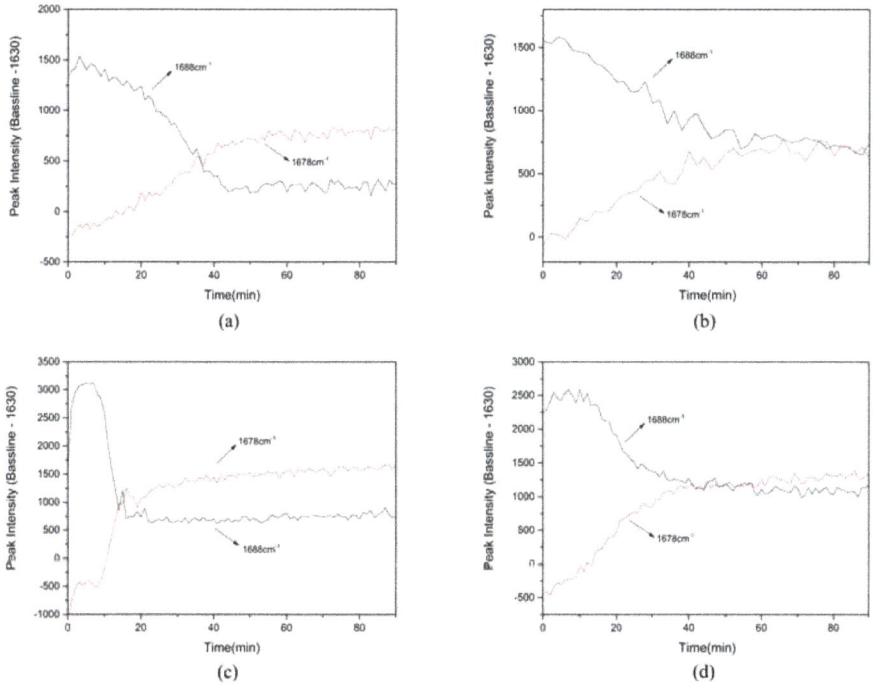

Figure 6. The relative peak intensity during TP-BA cocrystal formation in slurry crystallization: (a) exp 1; (b) exp 2; (c) exp 3; (d) exp 4.

Comparing the cocrystal formation time under different conditions in slurry crystallization, we found that the suspension density of raw materials and the temperature both have impact on the transformation from raw materials to cocrystal. From the results of exp 1 and 2 (Table 3, Figure 6), the cocrystal formation time of exp 1 is obviously shorter than that of exp 2, while the initial concentration of raw materials in exp 1 is higher than exp 2. The results show that when the initial concentration of raw materials is higher, TP-BA cocrystal formation is faster. The initial concentration directly affects the suspension density of raw materials, and higher suspension density can improve the collision probability of raw material particles in slurry crystallization to expedite cocrystal formation. Meanwhile, the total contact surface area between the API and coformer molecules in suspension can be increased as the suspension density becomes higher, and this can also improve cocrystal formation rate. Therefore, the suspension density of raw materials can affect cocrystal formation and the cocrystal formation rate would increase as the suspension density of raw materials increases. The same results can be achieved by comparing the transformation time from raw materials to cocrystal in exp 3 and 4. The lower formation time in exp 3 indicates higher suspension density in exp 3 (0.078 M) can achieve a higher rate of transformation from raw materials to TP-BA cocrystal than in exp 4 (suspension density is 0.042 M) when reaction temperature is same.

Comparing Raman spectra of exp 1 and 3, under the same suspension density condition, TP-BA cocrystal formation is faster in exp 3 when the temperature is higher than that of exp 1. This is mainly because higher temperature will facilitate the raw materials to reach the activated state with less time, and can thus shorten the formation time. The spectra of on-line monitoring of exp 2 and 4 show the same results. With the same suspension density of raw materials exp 4 (313.15 K) has a higher rate of TP-BA cocrystal formation than exp 2 (298.15 K), which also indicates temperature has an impact on TP-BA cocrystal formation under the same suspension density condition.

### 3.5. On-Line Monitoring of TP-BA Cocrystal Formation Process in Cooling Crystallization

Similar to the slurry experiments, Raman spectroscopy can be also used to monitor the cooling crystallization process of TP-BA cocrystal. In cooling crystallization, the TP-BA cocrystal was prepared in a methanol/water mixture (V:V = 1:5). The Raman spectra obtained from cooling crystallization, which has the same peak shifts as in slurry experiments, can provide various data of the cocrystallization process, including nucleation time, nucleation temperature and suitable cooling ending point.

Two couples of peaks at 1678/1688 $cm^{-1}$ and 918/925 $cm^{-1}$ are regarded as the characteristics peaks to investigate the cocrystal formation process in cooling crystallization. The corresponding relative peak intensity is calculated by the above method and shown in Figure 7. It can be seen from Figure 7 that the peak intensities of 1678 and 925 $cm^{-1}$, which represent the TP-BA cocrystal, start to increase at 170 min, and those of 1688 and 918 $cm^{-1}$, which represent the raw materials, begin decreasing at the same time. This means that at 170 min in the cooling process cocrystal nucleation occurs. Further, the nucleation temperature of TP-BA cocrystal can be calculated from the combined nucleation time and cooling rate, which is at 296.65 K. In theory, during the cooling process the amount of TP-BA cocrystal should keep increasing because the cocrystal solubility decreases continuously as temperature decreases. However, from the Raman spectra of the on-line monitoring of cooling crystallization (Figure 7), the relative peak intensity that represents the cocrystal shows no significant increase and stays at an approximately stable level after 250 min of the cooling process. That is because when cocrystal solubility decreases to a relatively low level, the yield cannot be increased significantly by further cooling. The corresponding temperature is the ending point of the cooling process. In the TP-BA cooling crystallization process, the ending temperature obtained from the Raman spectra of on-line monitoring is 284.65 K. Hence, on-line monitoring in cooling crystallization by Raman spectroscopy can be used to monitor cocrystal formation and to obtain information about the cocrystal nucleation temperature and ending temperature in cooling crystallization.

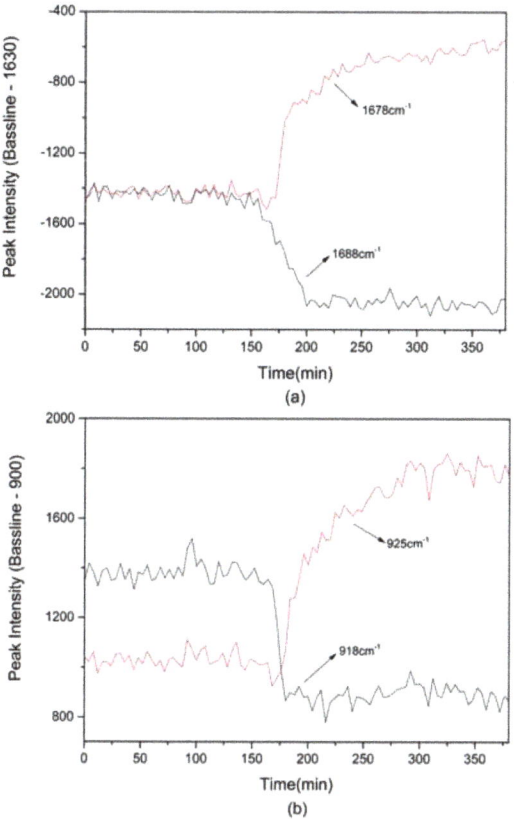

**Figure 7.** The relative peak intensity during TP-BA cocrystal formation in cooling crystallization: (a) 1678/1688 cm$^{-1}$; (b) 918/925 cm$^{-1}$.

## 4. Conclusions

In this work, the theophylline-benzoic acid cocrystal was prepared in solution via slurry and cooling crystallization. The solution synthesis of TP-BA cocrystal overcomes the disadvantages of existing preparation methods and can be used to produce large scale and high purity TP-BA cocrystal. The TP-BA cocrystal was characterized by PXRD, DSC, TGA and Raman spectroscopy. The cocrystallization process of TP and BA in solution was monitored by on-line Raman spectroscopy (both slurry and cooling crystallization). The change of peak intensity in the cocrystal formation process can be observed easily from the Raman spectra. The factors affecting the cocrystal formation rate in slurry crystallization were also explored. The results of Raman spectra show that suspension density of drug and coformer, and temperature, both have an impact on TP-BA cocrystal formation. The cocrystallization process will become faster as the suspension density increases. When suspension density of the raw materials is the same, a higher temperature can improve the TP-BA cocrystal formation rate. Temperature and suspension density of raw materials both have a positive correlation with the TP-BA cocrystal formation rate. Furthermore, nucleation temperature and cooling ending point can be obtained by on-line monitoring of the cocrystal formation process in cooling crystallization.

**Supplementary Materials:** The following are available online at http://www.mdpi.com/2073-4352/9/7/329/s1, Figure S1: (a) The variation of Raman spectra at 1270–1360 cm$^{-1}$; (b) The change of relative peak intensity of 1323/1331 cm 1 in TP BA cocrystal formation in slurry crystallization at 298.15 K. Figure S2: (a) The variation of

Raman spectra at 1120–1200 cm$^{-1}$; (b) The change of relative peak intensity of 1161/1171 cm$^{-1}$ in TP-BA cocrystal formation in slurry crystallization at 298.15 K. Figure S3: (a) The variation of Raman spectra at 890–940 cm$^{-1}$; (b) The change of relative peak intensity of 918/925 cm$^{-1}$ in TP-BA cocrystal formation in slurry crystallization at 298.15 K. Figure S4: HPLC chromatograms of slurry crystallization products in Table 3: exp 1 (a); exp 2 (b); exp 3 (c); exp 4 (d) and cooling crystallization product (e).Table S1: Information about Raman spectra in the range of 1640–1720 cm$^{-1}$ in Figure 5a. Table S2: Information about Raman spectra in the range of 1270–1360 cm$^{-1}$ in Figure S1a. Table S3: Information about Raman spectra in the range of 1120–1200 cm$^{-1}$ in Figure S2a. Table S4: Information about Raman spectra in the range of 890–940 cm$^{-1}$ in Figure S3a.

**Author Contributions:** Conceptualization, Y.H., L.Z. and Q.Y.; Formal analysis, W.Y., Y.Y. and C.W.; Funding acquisition, L.Z. and Q.Y.; Investigation, Y.H. and Q.Y.; Methodology, Y.H. and Q.Y.; Project administration, Y.H. and Q.Y.; Supervision, L.Z. and Q.Y.; Writing—original draft, Y.H.; Writing—review & editing, Y.H., Y.L., Z.Z. and X.Z.

**Funding:** The support from National Engineering Research Center for Industrial Crystallization Technology (NERCICT) is acknowledged. This work was supported by Major National Scientific Instrument Development Project of China (No. 21527812) and Tianjin Municipal Natural Science Foundation (No. 16JCZDJC32700).

**Conflicts of Interest:** The authors declare no competing financial interest.

### References

1. Byrn, S.R.; Zografi, G.; Chen, X. Accelerating proof of concept for small molecule drugs using solid-state chemistry. *J. Pharm. Sci.* **2010**, *99*, 3665–3675. [CrossRef] [PubMed]
2. Schultheiss, N.; Newman, A. Pharmaceutical cocrystals and their physicochemical properties. *Cryst. Growth Des.* **2009**, *9*, 2950–2967. [CrossRef] [PubMed]
3. Aakeröy, C.B.; Salmon, D.J. Building co-crystals with molecular sense and supramolecular sensibility. *CrystEngComm* **2005**, *7*, 439–448. [CrossRef]
4. Rodríguez-Hornedo, N. Cocrystals: Molecular design of pharmaceutical materials. *Mol. Pharm.* **2007**, *4*, 299–300. [CrossRef]
5. Liao, X.M.; Gautam, M.; Grill, A.; Zhu, H.J.J. Effect of position isomerism on the formation and physicochemical properties of pharmaceutical co-crystals. *J. Pharm. Sci.* **2010**, *99*, 246–254. [CrossRef] [PubMed]
6. Thanigaiman, K.; Khalib, N.C.; Temel, E.; Arshad, S.; Razak, I.A. New supramolecular cocrystal of 2-amino-5-chloropyridine with 3-methylbenzoic acids: Syntheses, structural characterization, hirshfeld surfaces and quantum chemical investigations. *J. Mol. Struct.* **2015**, *1099*, 246–256. [CrossRef]
7. Hickey, M.B.; Peterson, M.L.; Scoppettuolo, L.A.; Morrisette, S.L.; Vetter, A.; Guzmán, H.; Remenar, J.F.; Zhang, Z.; Tawa, M.D.; Haley, S.; et al. Performance comparison of a co-crystal of carbamazepine with marketed product. *Eur. J. Pharm. Biopharm.* **2007**, *67*, 112–119. [CrossRef]
8. Remenar, J.F.; Perterson, M.L.; Stephens, P.W.; Zhang, Z.; Zimenkov, Y.; Hickey, M.B. Celecoxib: Nicotinamide dissociation: Using excipients to capture the cocrystal's potential. *Mol. Pharm.* **2007**, *4*, 386–400. [CrossRef]
9. Zhang, S.; Rasmuson, Å.C. The theophylline–oxalic acid co-crystal system: Solid phases, thermodynamics and crystallization. *CrystEngComm* **2012**, *14*, 4644–4655. [CrossRef]
10. Kulla, H.; Greiser, S.; Benemann, S.; Rademann, K.; Emmerling, F. In situ investigation of a self-accelerated cocrystal formation by grinding pyrazinamide with oxalic acid. *Molecules* **2016**, *21*, 917. [CrossRef]
11. Basavoju, S.; Boström, D.; Velaga, S.P. Indomethacin-saccharin cocrystal: Design, synthesis and preliminary pharmaceutical characterization. *Pharm. Res.* **2008**, *25*, 530–541. [CrossRef]
12. Chieng, N.; Rades, T.; Aaltonen, J. An overview of recent studies on the analysis of pharmaceutical polymorphs. *J. Pharm. Biomed. Anal.* **2011**, *55*, 618–644. [CrossRef]
13. Holaň, J.; Štěpánek, F.; Billot, P.; Ridvan, L. The construction, prediction and measurement of co-crystal ternary phase diagrams as a tool for solvent selection. *Eur. J. Pharm. Sci.* **2014**, *63*, 124–131. [CrossRef]
14. Ueto, T.; Takata, N.; Muroyama, N.; Nedu, A.; Sasaki, A.; Tanida, S.; Terada, K. Polymorphs and a hydrate of furosemide-nicotinamide 1:1 cocrystal. *Cryst. Growth Des.* **2012**, *12*, 485–494. [CrossRef]
15. Friščić, T.; Jones, W. Recent Advances in Understanding the Mechanism of Cocrystal Formation via Grinding. *Cryst. Growth Des.* **2009**, *9*, 1621–1637. [CrossRef]
16. Soares, F.L.F.; Carneiro, R.L. Green Synthesis of Ibuprofen–Nicotinamide Cocrystals and In-Line Evaluation by Raman Spectroscopy. *Cryst. Growth Des.* **2013**, *13*, 1510–1517. [CrossRef]

17. Lee, K.S.; Kim, K.J.; Ulrich, J. In Situ Monitoring of Cocrystallization of Salicylic Acid–4,4′-Dipyridyl in Solution Using Raman Spectroscopy. *Cryst. Growth Des.* **2014**, *14*, 2893–2899. [CrossRef]
18. Tong, Y.; Zhang, P.; Dang, L.; Wei, H. Monitoring of cocrystallization of ethenzamide saccharin: Insight into kineticprocess by in situ Raman spectroscopy. *Chem. Eng. Res. Des.* **2016**, *109*, 249–257. [CrossRef]
19. Kojima, T.; Tsutsumi, S.; Yamamoto, K.; Ikeda, Y.; Moriwaki, T. High-throughput cocrystal slurry screening by use of in situ Raman microscopy and multi-well plate. *Int. J. Pharm.* **2010**, *399*, 52–59. [CrossRef]
20. Trask, A.V.; Motherwell, W.D.S.; Jones, W. Physical stability enhancement of theophylline via cocrystallization. *Int. J. Pharm.* **2006**, *320*, 114–123. [CrossRef]
21. Abourahma, H.; Urban, J.M.; Morozowich, N.; Chan, B. Examining the robustness of a theophylline cocrystal during grinding with additives. *CrystEngComm* **2014**, *14*, 6163–6169. [CrossRef]
22. Alhalaweh, A.; Kaialy, W.; Buckton, G.; Gill, H.; Nokhodchi, A.; Velaga, S.P. Theophylline Cocrystals Prepared by Spray Drying: Physicochemical Properties and Aerosolization Performance. *AAPS PharmSciTech.* **2013**, *14*, 265–276. [CrossRef]
23. Fulias, A.; Soica, C.; Ledeti, I.; Vlase, T.; Vlase, G.; Suta, L.M.; Belu, I. Characterization of Pharmaceutical Acetylsalicylic Acid-theophylline Cocrystal Obtained by Slurry Method Under Microwave Irradiation. *Rev. Chim.* **2014**, *65*, 1281–1284.
24. Lin, H.L.; Hsu, P.C.; Lin, S.Y. Theophylline-citric acid co-crystals easily induced by DSC-FTIR microspectroscopy or different storage conditions. *Asian J. Pharm. Sci.* **2013**, *8*, 18–26. [CrossRef]
25. Lu, J.; Rohani, S. Preparation and Characterization of Theophylline-Nicotinamide Cocrystal. *Org. Process Res. Dev.* **2009**, *13*, 1269–1275. [CrossRef]
26. Zhang, S.; Chen, H.; Rasmuson, Å.C. Thermodynamics and crystallization of a theophylline-salicylic acid cocrystal. *CrystEngComm* **2015**, *17*, 4125–4135. [CrossRef]
27. Heiden, S.; Tröbs, L.; Wenzel, K.J.; Emmerling, F. Mechanochemical synthesis and structural characterisation of a theophylline-benzoic acid cocrystal (1:1). *CrystEngComm* **2012**, *14*, 5128–5129. [CrossRef]
28. Zhang, S.; Rasmuson, Å.C. Thermodynamics and Crystallization of the Theophylline-Glutaric Acid Cocrystal. *Cryst. Growth Des.* **2013**, *13*, 1153–1161. [CrossRef]
29. Widhalm, J.R.; Dudareva, N. A familiar ring to it: Biosynthesis of plant benzoic acids. *Mol. Plant.* **2015**, *8*, 83–97. [CrossRef]
30. Childs, S.L.; Stahly, G.P.; Park, A. The Salt-Cocrystal Continuum: The Influence of Crystal Structure on Ionization State. *Mol. Pharm.* **2007**, *4*, 323–338. [CrossRef]
31. Sheikh, A.Y.; Rahim, S.A.; Hammond, R.B.; Roberts, K.J. Scalable solution cocrystallization: Case of carbamazepine-nicotinamide I. *CrystEngComm* **2009**, *11*, 501–509. [CrossRef]

 © 2019 by the authors. Licensee MDPI, Basel, Switzerland. This article is an open access article distributed under the terms and conditions of the Creative Commons Attribution (CC BY) license (http://creativecommons.org/licenses/by/4.0/).

Article

# Melting Diagrams of Adefovir Dipivoxil and Dicarboxylic Acids: An Approach to Assess Cocrystal Compositions

Hyunseon An, Insil Choi and Il Won Kim *

Department of Chemical Engineering, Soongsil University, Seoul 06978, Republic of Korea; hs4558@soongsil.ac.kr (H.A.); cisq@nate.com (I.C.)
* Correspondence: iwkim@ssu.ac.kr; Tel.: +82-2-820-0614; Fax.: +82-2-812-5378

Received: 30 December 2018; Accepted: 30 January 2019; Published: 30 January 2019

**Abstract:** Pharmaceutical cocrystallization is a useful method to regulate the physical properties of active pharmaceutical ingredients (APIs). Since the cocrystals may form in various API/coformer ratios, identification of the cocrystal composition is the critical first step of any further analysis. However, the composition identification is not always unambiguous if cocrystallization is performed in solid state with unsuccessful solution crystallization. Single melting point and some new X-ray diffraction peaks are necessary but not sufficient conditions. In the present study, the use of melting diagrams coupled with the X-ray diffraction data was tested to identify cocrystal compositions. Adefovir dipivoxil (AD) was used as a model API, and succinic acid (SUC), suberic acid (SUB), and glutaric acid (GLU) were coformers. Compositions of AD/SUC and AD/SUB had been previously identified as 2:1 and 1:1, but that of AD/GLU was not unambiguously identified because of the difficulty of solution crystallization. Melting diagrams were constructed with differential scanning calorimetry, and their interpretation was assisted by powder X-ray diffraction. The cocrystal formation was exhibited as new compositions with congruent melting in the phase diagrams. This method correctly indicated the previously known cocrystal compositions of AD/SUC and AD/SUB, and it successfully identified the AD/GLU cocrystal composition as 1:1. The current approach is a simple and useful method to assess the cocrystal compositions when the crystallization is only possible in solid state.

**Keywords:** pharmaceutical cocrystal; melting diagram; liquid assisted grinding; adefovir dipivoxil; dicarboxylic acid

---

## 1. Introduction

Pharmaceutical crystallization plays a critical role in solid dosage forms, because the physicochemical properties of the solid forms of active pharmaceutical ingredients (APIs) are intrinsically defined by the structures of the API crystals. Salt formation is perhaps the most well-known example, and polymorphs, solvates, hydrates, and cocrystals are the important variations of the solid forms [1,2].

Pharmaceutical cocrystallization is an emerging technology that involves strong intermolecular interactions (usually hydrogen bonding) between APIs and coformers [3–5]. The diverse possibilities of coformers have expanded the landscape of pharmaceutical crystallization in a new dimension. The utility of the cocrystal formation includes the improvements of solubility, stability, processability, and so on [6–9]. Some examples of pharmaceutical cocrystals commercially available on the market are Farxiga™, Suglat™, Steglatro™, and Entresto™, and the cases are expected to increase steadily [10,11]. Experimental screening methods for the viable pairs of APIs/coformers are diverse, and a single method does not consider the entire range of cocrystal possibilities. Solution crystallization with cooling or solvent evaporation is perhaps most in line with the traditional crystallization processes [12,13]. However, spray drying is sometimes effective to discover the cocrystals not easily found using the usual solution method [14]. Liquid-assisted grinding is another versatile method bypassing solution crystallization, and it requires a relatively small amount of samples [3,15]. In addition, melt screening with an API/coformer mixing zone can effectively cover the diverse range of API/coformer compositions, which is at least qualitatively equivalent to the binary phase diagram [16].

Ultimately, the nature of API–coformer interactions as well as structural conformations is necessary to completely understand the characteristics of cocrystals. The usual method for the full characterization is single crystal X-ray diffraction (XRD), where the growth of relatively large crystals is required. Unfortunately, the formation of suitable single crystals is often elusive, and cocrystal generation through solution crystallization is not always straightforward [9,17,18]. In these cases, an alternative route is the solid-state preparation of cocrystals followed by the analysis of high-quality powder XRD data assisted by the Rietveld method, Monte Carlo simulated annealing, and molecular dynamics [19–21]. For this, the first imperative step is the identification of the correct API–coformer composition. Incorrect API/coformer mixtures during solid-state preparation may generate cocrystals contaminated by excess starting materials (i.e., API or coformer), which would in turn make it impossible to acquire high quality powder X-ray data suitable for structural analysis. (Note that the usual quantitative analysis of the solution (e.g., nuclear magnetic resonance spectroscopy) cannot be applied for the stoichiometry determination without pure cocrystals, which requires the knowledge of the predetermined stoichiometry in the case of the solid-state preparation.).

In the present study, we attempted to determine the API/coformer compositions of cocrystals through the combination of melting diagrams and powder XRD data of the API/coformer mixtures [22]. Adefovir dipivoxil (AD) with dicarboxylic acids (Figure 1: succinic (SUC), glutaric (GLU), and suberic (SUB) acids), long studied in our research group, was employed [23–26], and the mixtures were prepared by liquid-assisted grinding. We note here that the compositions of the adefovir dipivoxil cocrystals with succinic acid or suberic acid as coformers were identified undoubtedly in our previous studies of single crystal XRD [23,24], whereas that with glutaric acid was not unambiguously determined because of the difficulty of growing crystals through solution crystallization. (We re-emphasize that such difficulty has been shared in some other API cocrystal systems [9,17,18].) The evidence in our previous study with glutaric acid, a single melting point and some new XRD peaks [26], turned out to be necessary but not sufficient conditions to make clear conclusions about the cocrystal composition as is shown in the present study.

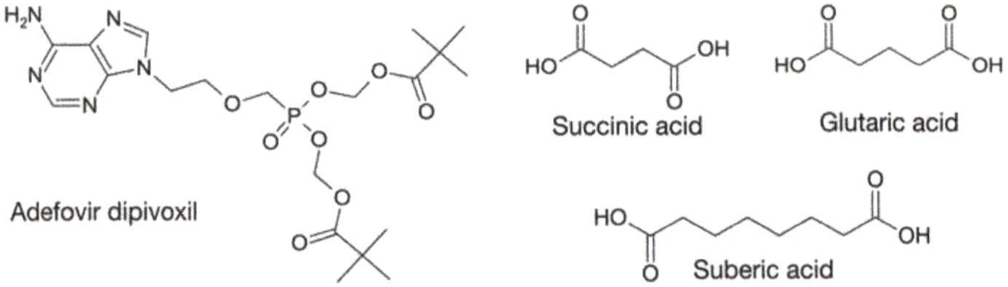

**Figure 1.** Chemical structures: adefovir dipivoxil (AD); succinic acid (SUC); glutaric acid (GLU); suberic acid (SUB).

## 2. Materials and Methods

*2.1. Cocrystallization*

Cocrystallization of adefovir dipivoxil (AD, form I: ≥98.0%, Tokyo Chemical Industry Co., Tokyo, Japan; ≥99.68%, Carbosynth, Berkshire, UK) and dicarboxylic acids was performed through liquid-assisted grinding at room temperature (25–26 °C) with an agate mortar and pestle. Dicarboxylic acids were succinic acid (SUC, ≥99.0%), glutaric acid (GLU, 99%), and suberic acid (SUB, 98%), and they were all purchased from Sigma-Aldrich (Milwaukee, WI, USA). Chemical structures of the compounds are shown in Figure 1.

All experiments were at 0.2 mmol (approximately 100 mg) AD scale, and various molar ratios of AD/dicarboxylic acid pairs were ground for 30 min with the addition of methanol (HPLC grade, J.T. Baker, Phillipsburg, NJ, USA) or ethanol (anhydrous 99.9%, Samchun Chemical, Seoul, South Korea). For the AD/SUC mixture, 40 µL ethanol was added every 5 min during grinding. For the AD/GLU mixture, 40 µL methanol was added every 10 min. For AD/SUB, 40 µL methanol was added every 5 min. After grinding, each product was moved to an open 4-mL glass vial to dry for 2 h at 35–45 °C in a vacuum oven (J-DVO1, Jisico, Seoul, South Korea).

## 2.2. Characterization

Thermal properties of the ground AD/dicarboxylic acids were investigated in various molar ratios using a differential scanning calorimeter (DSC: DSC 812e, Mettler-Toledo, Columbus, OH, USA). DSC was pre-calibrated with indium for temperature and enthalpy. Temperature scan was from 25 to 200 °C with a scanning rate of 10 °C/min under $N_2$ (50 mL/min). A typical sample amount in an Al crucible (40 µL volume with a top pinhole) was 2–4 mg. Melting points were measured at the onset temperatures of the first melting endotherms and at the peak temperatures of the final melting endotherms because the final endotherms were sometimes broad or overlapped.

Crystal phases were identified via X-ray diffraction (XRD: D2 PHASER, Bruker AXS, Billerica, MA, USA). The 2θ-θ mode was employed to measure 2θ range 6–30° (1°/min, increment 0.02°) with CuKα radiation ($\lambda$ = 1.5406 Å) at 30 kV and 10 mA. A Si low-background sample holder (Bruker AXS, Billerica, MA, USA) was used for increased sensitivity.

## 3. Results and Discussion

AD cocrystals with SUC and SUB were studied to evaluate the effectiveness of using the binary phase diagrams to assess the cocrystal compositions. The phase diagrams of AD/SUC and AD/SUB were constructed from the melting behavior of the powders, observed via DSC, which were obtained through liquid-assisted grinding. In addition, the crystal phases were identified using XRD.

The phase diagram of AD and SUC is shown in Figure 2a. Single melting behavior was observed at four different compositions: pure SUC ($x_{AD}$ = 0, where x is mole fraction), AD/SUC = 2:3 and 2:1, and pure AD ($x_{AD}$ = 1), in the order of increasing AD contents (Figures 2a,b and S1). AD/SUC = 2:3 ($x_{AD}$ = 0.4) was the eutectic composition formed of SUC ($x_{AD}$ = 0) and AD/SUC = 2:1 ($x_{AD} \cong 0.67$). When $x_{AD}$ < 0.4, double melting behavior was observed: the eutectic melting and the melting of extra SUC of which the melting point was depressed proportional to the AD addition. When $x_{AD}$ > 0.4 in the AD/SUC mixture, single melting was observed only at AD/SUC = 2:1. Other compositions exhibited either eutectic melting or extra AD melting before reaching the other melting point, which was lower than the single melting point observed at AD/SUC = 2:1 (Figure 2). This indicates that the higher melting points when $x_{AD}$ > 0.4 are the depressed melting points of AD/SUC cocrystals due to extra SUC (0.4 < $x_{AD}$ < 0.67) or extra AD (0.67 < $x_{AD}$). The observation of the melting behavior at diverse compositions revealed that the cocrystal of AD/SUC = 2:1 formed during the liquid-assisted grinding (onset temperature 123 °C; heat of fusion 109 J/g).

XRD (Figures 2c and S2) also supported the conclusion made from the interpretation of the phase diagram. Representative XRD patterns are shown in Figure 2c. At AD/SUC = 2:1, new diffraction peaks appeared (e.g., 2θ = 9.46, 14.82, and 23.93°, marked by star), which were absent in neat AD and SUC; characteristic AD (inverse triangle) and SUC (triangle) peaks disappeared as well, indicating complete conversion of AD and SUC into the new solid phase. At AD/SUC = 2:3, while the new diffraction peaks (star) showed up, residual SUC peak (triangle) remained, revealing that the extra SUC phase was present in addition to the new cocrystal phase.

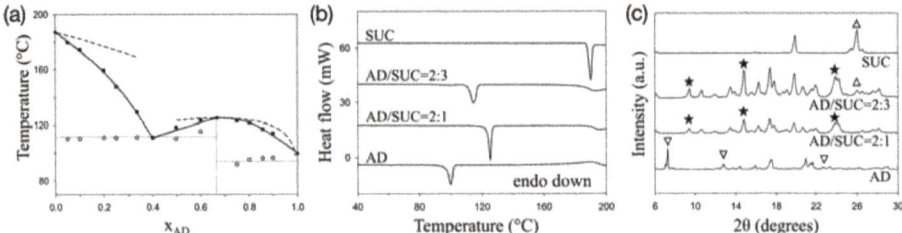

**Figure 2.** AD/SUC system: (**a**) melting diagram; (**b**) representative DSC thermograms; (**c**) representative XRD patterns. Top dashed lines in the melting diagram represent ideal liquidus lines. Some X-ray diffractions are marked as stars, triangles, and inverse triangles for cocrystal, SUC, and AD, respectively.

Overall, the aforementioned observations allowed us to conclude that the AD/SUC cocrystal composition was 2:1. This is in agreement with the previous single crystal XRD study, independently confirming the cocrystal composition [23]. We note that the liquid-assisted grinding of AD/SUC was performed with the addition of ethanol; methanol induced the polymorphism of the AD/SUC cocrystal, which significantly complicated the XRD and DSC analysis of the powder mixtures [25].

We also noted evidence suggesting partial miscibility between the AD/SUC cocrystal and SUC. The experimental heat of fusion at the eutectic composition of AD/SUC = 2:3 was 95 J/g, whereas the theoretical value was 146 J/g based on the heat of fusion for neat cocrystal (109 J/g) and SUC (321 J/g) (Figure 2b). In addition, XRD peaks of the cocrystal at 21.54° and 22.01° displayed subtle changes (Figure S2). As the amount of extra SUC increased (from AD/SUC = 2:1 to 1:19), the 22.01° peak was intensified and shifted to 21.86°, and the 21.54° peak disappeared (no SUC diffraction peak exists in this diffraction region).

The AD/SUB system showed very similar behavior to the AD/SUC case, except that the eutectic and cocrystal formations were at different compositions. Single melting behavior was observed at four different compositions: pure SUB ($x_{AD}$ = 0), AD/SUB = 1:3 and 1:1, and pure AD ($x_{AD}$ = 1), in the order of increasing AD contents (Figures 3a,b and S3). (Phase transformation of SUB itself at 131 °C was neglected for the purpose of diagram construction, and only the final melting was considered [27].) AD/SUB = 1:3 was the eutectic composition formed of SUB ($x_{AD}$ = 0) and AD/SUB = 1:1 ($x_{AD}$ = 0.5). When $x_{AD}$ < 0.25, double melting behavior was observed: the eutectic melting and the melting of extra SUB of which the melting point was depressed proportional to the AD addition. When $x_{AD}$ > 0.25 in the AD/SUB mixture, single melting was observed only at AD/SUB = 1:1. Other compositions exhibited either eutectic melting or extra AD melting before reaching the other melting point, which was lower than the single melting point observed at AD/SUB = 1:1 (Figure 3). This indicates that the higher melting points when $x_{AD}$ > 0.25 are the depressed melting points of AD/SUB cocrystals due to extra SUB (0.25 < $x_{AD}$ < 0.5) or extra AD (0.5 < $x_{AD}$). The observation of the melting behavior at diverse compositions revealed that the cocrystal of AD/SUB = 1:1 formed during the liquid-assisted grinding (onset temperature 131 °C; heat of fusion 148 J/g). We note that the liquid-assisted grinding of AD/SUB was performed with the addition of methanol, and methanol solvate appeared with the excess AD ($x_{AD}$ > 0.5). The solvate-related data could be easily identifiable (melting point at around 79 °C) [28], and solvate domain in the phase diagram was omitted for simplification.

XRD (Figures 3c and S4) also supported the interpretation of the phase diagram. Representative XRD patterns are shown in Figure 3c. At AD/SUB = 1:1, new diffraction peaks appeared (e.g., $2\theta$ = 7.01, 9.07, 18.54, 19.51°, marked by star), which were absent in neat AD and SUB; characteristic AD (inverse triangle) and SUB (triangle) peaks disappeared as well, indicating complete conversion of AD and SUB into the new solid phase. At AD/SUB = 1:3, while the new diffraction peaks (star) showed up, residual SUB peak (triangle) remained, revealing that the extra SUB phase was present in addition to the new cocrystal phase.

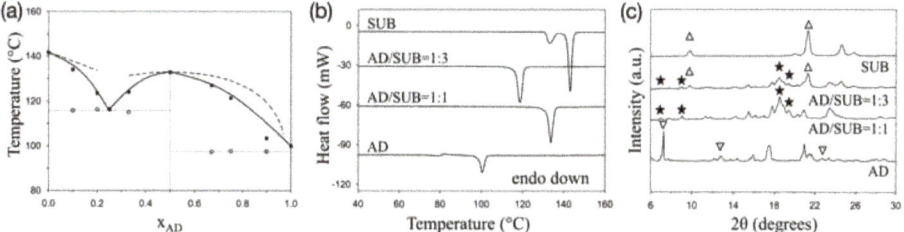

**Figure 3.** AD/SUB system: (**a**) melting diagram; (**b**) representative DSC thermograms; (**c**) representative XRD patterns. Top dashed lines in the melting diagram represent ideal liquidus lines. Some X-ray diffractions are marked as stars, triangles, and inverse triangles for cocrystal, SUB, and AD, respectively.

Overall, these observations allowed us to conclude that the AD/SUB cocrystal composition was 1:1. This is in agreement with the previous single crystal XRD study, independently confirming the cocrystal composition [24]. From the studies of AD/SUC and AD/SUB systems, the approach of constructing the phase diagram complemented by XRD analysis was hitherto proved successful for identifying the correct cocrystal compositions.

The AD/GLU system was investigated using the same approach, of which cocrystal composition had not been unambiguously determined because of the difficulty of cocrystal formation from solutions. Figure 4a shows the melting diagram of AD and GLU. Single melting behavior was observed at five different compositions: pure GLU ($x_{AD} = 0$), AD/GLU = 1:3, 1:1, and 2:1, and pure AD ($x_{AD} = 1$), in the order of increasing AD contents (Figures 4a,b and S5). (Phase transformation of GLU itself at 75 °C was neglected for the purpose of diagram construction, and only the final melting was considered [29].) AD/GLU = 1:3 was the eutectic composition formed of GLU ($x_{AD} = 0$) and AD/GLU = 1:1 ($x_{AD} = 0.5$). Similarly, AD/GLU = 2:1 ($x_{AD} \cong 0.67$) was the eutectic composition formed of AD/GLU = 1:1 ($x_{AD} = 0.5$) and AD ($x_{AD} = 1$). The compositions around the eutectic points exhibited the eutectic melting as well as the depressed melting of the phases involved in the eutectic formation, namely, neat GLU, neat AD, and AD/GLU = 1:1. The observation of the melting behavior at diverse compositions made us conclude that the cocrystal of AD/GLU = 1:1 formed during the liquid-assisted grinding (onset temperature 87 °C; heat of fusion 83 J/g).

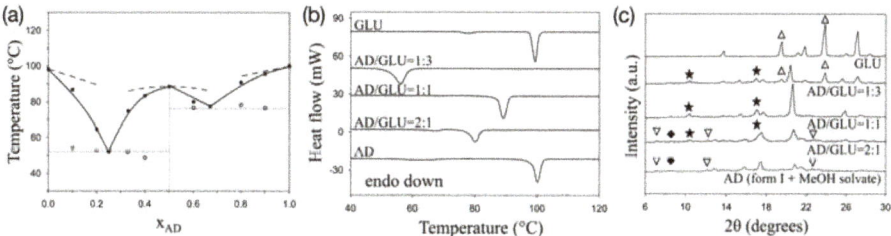

**Figure 4.** AD/GLU system: (**a**) melting diagram; (**b**) representative DSC thermograms; (**c**) representative XRD patterns. Top dashed lines in the melting diagram represent ideal liquidus lines. Some X-ray diffractions are marked as stars, triangles, inverse triangles, and diamonds for cocrystal, GLU, AD, and AD methanol solvate, respectively.

XRD (Figures 4c and S6) also supported the interpretation of the phase diagram. Representative XRD patterns are shown in Figure 4c. At AD/GLU = 1:1, new diffraction peaks appeared (e.g., $2\theta$ = 10.42 and 17.67°, marked by star), which were absent in neat AD and GLU; characteristic AD (inverse triangle) and GLU (triangle) peaks disappeared as well, indicating complete conversion of AD and GLU into the new solid phase. At AD/GLU = 1:3, while the new diffraction peaks (star) showed up, residual GLU peaks (triangle) remained, implying that the extra GLU phase was present in addition to the new cocrystal phase. At AD/GLU = 2:1, while the new diffraction peaks (star) showed up, residual AD peaks (inverse triangle) remained, suggesting that the extra AD phase was present in addition to the new cocrystal phase.

Overall, the melting diagram approach assisted by XRD allowed us to conclude that the previously unknown composition of AD/GLU cocrystal was 1:1. We note that the liquid-assisted grinding of AD/GLU was performed with the addition of methanol because of the too slow evaporation of ethanol during the grinding process, and the methanol solvate-related domain in the phase diagram (when $x_{AD} > 0.5$) was omitted for clarity.

We also noted evidence suggesting partial miscibility between the AD/GLU cocrystal and GLU. An XRD peak of the cocrystal at 20.62° displayed subtle changes when extra GLU existed (Figure 4c and Figure S6). This peak experienced a downward shift with extra GLU, and the shift was greatest to 20.42° at the eutectic composition (AD/GLU = 1:3). Eutectic analysis based on the heat of fusion was unfortunately not possible. Since the eutectic temperature was below the phase transformation temperature of neat GLU, the theoretical heat of fusion during eutectic melting could not be calculated.

The emphasis of the current study was to find definite compositions with congruent melting using the phase diagrams, which were equivalent to the cocrystal compositions [30]. Further studies on the phase diagrams, for example, deviations from the ideal liquidus lines based on equilibrium theories (top dashed lines in Figures 2a, 3a, and 4a) and partial miscibility, may shed light on the formation mechanisms of the cocrystals and their compositional diversity [30,31].

## 4. Conclusions

Melting diagrams of AD and dicarboxylic acids were utilized to identify the compositions of AD/dicarboxylic acid cocrystals. DSC was employed for the observation of thermal behavior at different ratios of AD/coformer mixtures, and XRD was used to identify the solid phases of the powder mixtures. AD/SUC = 2:1 and AD/SUB = 1:1, which had been previously determined through single crystal XRD, were correctly identified through the current phase diagram approach. Moreover, the composition of AD/GLU cocrystal was newly identified as 1:1, which could not be previously specified due to the difficulty of solution crystal growth. The current approach can be summarized as finding new compositions with congruent melting in the melting diagrams. It will be useful for the cocrystals that cannot be easily grown using the solution method, since the identification of the cocrystal compositions is the critical first step toward the more quantitative structural analysis.

**Supplementary Materials:** The following are available online at http://www.mdpi.com/2073-4352/9/2/70/s1, Figure S1: DSC thermograms for the AD/SUC system, Figure S2: XRD patterns for the AD/SUC system, Figure S3: DSC thermograms for the AD/SUB system, Figure S4: XRD patterns for the AD/SUB system, Figure S5: DSC thermograms for the AD/GLU system, Figure S6: XRD patterns for the AD/GLU system.

**Author Contributions:** I.W.K and H.A. designed the experiments, analyzed the data, and wrote the paper; H.A. performed most of the experiments; I.C. helped H.A. with some of the experiments.

**Acknowledgments:** This research was supported by the Basic Science Research Program through the National Research Foundation of Korea (NRF) funded by the Ministry of Education (NRF- 2015R1D1A1A01058116).

**Conflicts of Interest:** The authors declare no conflict of interest. The funding sponsors had no role in the design of the study; in the collection, analyses, or interpretation of data; in the writing of the manuscript; and in the decision to publish the results.

## References

1. Stahl, P.H.; Nakano, M. Pharmaceutical aspects of the drug salt form. In *Handbook of Pharmaceutical Salts*, 1st ed.; Stahl, P.H., Wermuth, C.G., Eds.; WILEY-VCH: Weinheim, Germany, 2008; pp. 83–116. ISBN 978-3-906390-58-1.
2. Rao, V.M.; Sanghvi, R.; Zhu, H. Solubility of pharmaceutical solids. In *Developing Solid Oral Dosage Forms*, 1st ed.; Qiu, Y., Chen, Y., Zhang, G.G.Z., Liu, L., Porter, W.R., Eds.; Academic Press: Burlington, MA, USA, 2009; pp. 1–24. ISBN 978-0-444-53242-8.

3. Jones, W.; Motherwell, W.D.S.; Trask, A.V. Pharmaceutical cocrystals: An emerging approach to physical property enhancement. *MRS Bull.* **2006**, *31*, 875–879. [CrossRef]
4. Shan, N.; Zaworotko, M.J. The role of cocrystals in pharmaceutical science. *Drug Discov. Today* **2008**, *13*, 440–446. [CrossRef] [PubMed]
5. Rodríguez-Hornedo, N.; Nehm, S.J.; Jayasankar, A. Cocrytals: Design, properties and formation mechanisms. In *Encyclopedia of Pharmaceutical Technology*, 3rd ed.; Swarbrick, J., Ed.; Informa Healthcare: New York, NY, USA, 2007; Volume 1, pp. 615–635. ISBN 978-0-8493-9396-9.
6. Chow, S.F.; Chen, M.; Shi, L.; Chow, A.H.L.; Sun, C.C. Simultaneously improving the mechanical properties, dissolution performance, and hygroscopicity of ibuprofen and flurbiprofen by cocrystallization with nicotinamide. *Pharm. Res.* **2012**, *29*, 1854–1865. [CrossRef] [PubMed]
7. Good, D.J.; Rodríguez-Hornedo, N. Solubility advantage of pharmaceutical cocrystals. *Cryst. Growth Des.* **2009**, *9*, 2252–2264. [CrossRef]
8. Trask, A.V.; Motherwell, W.D.S.; Jones, W. Physical stability enhancement of theophylline via cocrystallization. *Int. J. Pharm.* **2006**, *320*, 114–123. [CrossRef] [PubMed]
9. Karki, S.; Friščić, T.; Fábián, L.; Laity, P.R.; Day, G.M.; Jones, W. Improving mechanical properties of crystalline solids by cocrystal formation: New compressible forms of paracetamol. *Adv. Mat.* **2009**, *21*, 3905–3909. [CrossRef]
10. Wood, P.A.; Feeder, N.; Furlow, M.; Galek, P.T.A.; Groom, C.R.; Pidcock, E. Knowledge-based approaches to co-crystal design. *CrystEngComm* **2014**, *16*, 5839–5848. [CrossRef]
11. Kumar, A.; Kumar, S.; Nanda, A. A review about regulatory status and recent patents of pharmaceutical co-crystals. *Adv. Pharm. Bull.* **2018**, *8*, 355–363. [CrossRef]
12. Yu, Z.Q.; Chow, P.S.; Tan, R.B.H. Operating regions in cooling cocrystallization of caffeine and glutaric acid in acetonitrile. *Cryst. Growth Des.* **2010**, *10*, 2382–2387. [CrossRef]
13. Zhang, S.; Chen, H.; Rasmuson, Å.C. Thermodynamics and crystallization of a theophylline–salicylic acid cocrystal. *CrystEngComm* **2015**, *17*, 4125–4135. [CrossRef]
14. Alhalaweh, A.; Kaialy, W.; Buckton, G.; Gill, H.; Nokhodchi, A.; Velaga, S.P. Theophylline cocrystals prepared by spray drying: Physicochemical properties and aerosolization performance. *AAPS PharmSciTech* **2013**, *14*, 265–276. [CrossRef] [PubMed]
15. Friščić, T.; Jones, W. Recent advances in understanding the mechanism of cocrystal formation via grinding. *Cryst. Growth Des.* **2009**, *9*, 1621–1637. [CrossRef]
16. Berry, D.J.; Seaton, C.C.; Clegg, W.; Harrington, R.W.; Coles, S.J.; Horton, P.N.; Hursthouse, M.B.; Storey, R.; Jones, W.; Friščić, T.; et al. Applying hot-stage microscopy to co-crystal screening: A study of nicotinamide with seven active pharmaceutical ingredients. *Cryst. Growth Des.* **2008**, *8*, 1697–1712. [CrossRef]
17. Shevchenko, A.; Bimbo, L.M.; Miroshnyk, I.; Haarala, J.; Jelínková, K.; Syrjänen, K.; van Veen, B.; Kiesvaara, J.; Santos, H.A.; Yliruusi, J. A new cocrystal and salts of itraconazole: Comparison of solid-state properties, stability and dissolution behavior. *Int. J. Pharm.* **2012**, *436*, 403–409. [CrossRef] [PubMed]
18. Karki, S.; Fábián, L.; Friščić, T.; Jones, W. Powder X-ray diffraction as an emerging method to structurally characterize organic solids. *Org. Lett.* **2007**, *9*, 3133–3136. [CrossRef] [PubMed]
19. Harris, K.D.M.; Tremayne, M.; Kariuki, B.M. Contemporary advances in the use of powder X-ray diffraction for structure determination. *Angew. Chem. Int. Ed.* **2001**, *40*, 1626–1651. [CrossRef]
20. Day, G.M.; van de Streek, J.; Bonnet, A.; Burley, J.C.; Jones, W.; Motherwell, W.D.S. Polymorphism of scyllo-inositol: Joining crystal structure prediction with experiment to elucidate the structures of two polymorphs. *Cryst. Growth Des.* **2006**, *6*, 2301–2307. [CrossRef]
21. Friščić, T.; Halasz, I.; Strobridge, F.C.; Dinnebier, R.E.; Stein, R.S.; Fábián, L.; Curfs, C. A rational approach to screen for hydrated forms of the pharmaceutical derivative magnesium naproxen using liquid-assisted grinding. *CrystEngComm* **2011**, *13*, 3125–3129. [CrossRef]
22. Évora, A.O.L.; Castro, R.A.E.; Maria, T.M.R.; Silva, M.R.; ter Horst, J.H.; Canotilho, J.; Eusébio, M.E.S. Co-crystals of diflunisal and isomeric pyridinecarboxamides—A thermodynamics and crystal engineering contribution. *CrystEngComm* **2016**, *18*, 4749–4759. [CrossRef]
23. Jung, S.; Ha, J.-M.; Kim, I.W. Bis[(2,2-dimethylpropanoyloxy)methyl]{[2-(6-amino-9H-purin-9-yl)ethoxy]methyl}phosphonate–succinic acid (2/1). *Acta Crystallogr. E* **2012**, *68*, o809–o810. [CrossRef] [PubMed]
24. Jung, S.; Lee, J.; Kim, I.W. Structures and physical properties of the cocrystals of adefovir dipivoxil with dicarboxylic acids. *J. Cryst. Growth* **2013**, *373*, 59–63. [CrossRef]

25. Jung, S.; Ha, J.-M.; Kim, I.W. Phase transformation of adefovir dipivoxil/succinic acid cocrystals regulated by polymeric additives. *Polymers* **2014**, *6*, 1–11. [CrossRef]
26. Jung, S.; Choi, I.; Kim, I.W. Liquid-assisted grinding to prepare a cocrystal of adefovir dipivoxil thermodynamically less stable than its neat phase. *Crystals* **2015**, *5*, 583–591. [CrossRef]
27. Roux, M.V.; Temprado, M.; Chickos, J.S. Vaporization, fusion and sublimation enthalpies of the dicarboxylic acids from C4 to C14 and C16. *J. Chem. Thermodyn.* **2005**, *37*, 941–953. [CrossRef]
28. Arimilli, M.N.; Kelly, D.E.; Lee, T.T.K.; Manes, L.V.; Munger, J.D., Jr.; Prisbe, E.J.; Schultze, L.M. Nucleotide Analog Compositions. U.S. Patent 6,451,340, 17 September 2002.
29. Dheep, G.R.; Sreekumar, A. Investigation on thermal reliability and corrosion characteristics of glutaric acid as an organic phase change material for solar thermal energy storage applications. *Appl. Therm. Eng.* **2018**, *129*, 1189–1196. [CrossRef]
30. Soustelle, M. *Phase Transformations*, 1st ed.; ISTE Ltd.: London, UK, 2015; pp. 75–112. ISBN 978-1-84821-868-0.
31. Prigogine, I.; Defay, R. *Chemical Thermodynamics*, 1st ed.; Longmans: London, UK, 1954; pp. 357–380. ISBN 978-0582462830.

© 2019 by the authors. Licensee MDPI, Basel, Switzerland. This article is an open access article distributed under the terms and conditions of the Creative Commons Attribution (CC BY) license (http://creativecommons.org/licenses/by/4.0/).

Article

# Synthesis, X-Ray Crystal Structure, Hirshfeld Surface Analysis, and Molecular Docking Study of Novel Hepatitis B (HBV) Inhibitor: 8-Fluoro-5-(4-fluorobenzyl)-3-(2-methoxybenzyl)-3,5-dihydro-4H-pyrimido[5,4-b]indol-4-one

Aleksandr V. Ivashchenko [1], Oleg D. Mitkin [1,*], Dmitry V. Kravchenko [2], Irina V. Kuznetsova [2], Sergiy M. Kovalenko [1,3], Natalya D. Bunyatyan [3,4] and Thierry Langer [5]

1. ChemRar Research and Development Institute, 7 Nobel st., Innovation Center Skolkovo Territory, 143026 Moscow, Russia
2. Chemical Diversity Research Institute, 2A Rabochaya st., Khimki, 141400 Moscow Region, Russia
3. Federal State Autonomous Educational Institution of Higher Education I.M. Sechenov First Moscow State Medical University of the Ministry of Healthcare of the Russian Federation (Sechenovskiy University). 8 Trubeckaya st., 119991 Moscow, Russia
4. Federal State Budgetary Institution "Scientific Centre for Expert Evaluation of Medicinal Products" of the Ministry of Health of the Russian Federation, Petrovsky boulevard 8, bld. 2, 127051 Moscow, Russia
5. Department of Pharmaceutical Chemistry, University of Vienna, Althanstraße 14, A-1090 Vienna, Austria
* Correspondence: mod.chemdiv@gmail.com; Tel.: +7-495-995-4941

Received: 6 June 2019; Accepted: 18 July 2019; Published: 24 July 2019

**Abstract:** A method for the synthesis of 8-fluoro-5-(4-fluorobenzyl)-3-(2-methoxybenzyl)-3,5-dihydro-4H-pyrimido[5,4-b]indol-4-one has been developed and the electronic and spatial structure of a new biologically active molecule has been studied both theoretically and experimentally. The title compound was crystallized from acetonitrile and the single-crystal X-ray analysis has revealed that it exists in a monoclinic P2$_1$/n space group, with one molecule in the asymmetric part of the unit cell, $a$ = 16.366(3) Å, $b$ = 6.0295(14) Å, $c$ = 21.358(4) Å, $\beta$ = 105.21(2)°, V = 2033.7(7) Å$^3$ and Z = 4. Hirshfeld surface analysis was used to study intermolecular interactions in the crystal. Molecular docking studies have evaluated the investigated compound as a new inhibitor of hepatitis B. Testing for anti-hepatitis B virus activity has shown that this substance has in vitro nanomolar inhibitory activity against Hepatitis B virus (HBV).

**Keywords:** hepatitis B; HBV; pharmaceutical crystals; 3,5-dihydro-4H-pyrimido[5,4-b]indol-4-one; 1H-indole; pyrimidin-4(3H)-one; hydrogen bond; Hirshfeld surface analysis; molecular docking study

---

## 1. Introduction

Despite significant progress in the etiology of viral hepatitis studying, the incidence of this disease remains quite high. Therefore, the problem of viral hepatitis is one of the most actual ones in modern medicine. According to the World Health Organization data, 240 million people worldwide suffer from chronic infection and are carriers of HBV. About 780 thousand people die each year from diseases associated with hepatitis B [1]. Currently, drugs to suppress viral replication are available, but a complete cure is rarely achieved. Therefore, the search for new drugs for the treatment of hepatitis B is an actual and important task.

Nitrogen-containing heterocycles play a key role among the multitude of biologically active heterocycles [2–6]. Recently, a literature search exposed that a large number of molecules currently under investigation contain nitrogen heterocycles, and of these, 1H-indoles and pyrimidin-4(3H)-ones form a very substantial type of compounds. These heterocycles have the "privileged" indole framework which is commonly found in natural products and pharmaceutical drugs [7–12]. Pyrimidoindole moieties [13–15] are important structural motifs which are found in numerous pharmaceutically active compounds. They have revealed a wide range of high biological activity, such as antihypertensive and anti-inflammatory [16], anti-asthma [17,18], and act as α1-adrenergic receptor ligands or A1 adenosine receptor antagonists [19], potential tyrosine kinases (PTK) inhibitors, CFR1 and neuropeptide Y receptor ligands [20]. Despite the large spectrum of biological activity of these moieties, this range can be supplemented by introducing certain functional groups. Therefore, it is of substantial interest to develop efficient methods for the synthesis of such structures and their derivatives.

In a continuation of our efforts to obtain new HBV inhibitors for treatment and prevention of human HBV infections [21,22], we initiated the design, synthesis, and anti-hepatitis B virus activity testing of a new 8-fluoro-5-(4-fluorobenzyl)-3-(2-methoxybenzyl)-3,5-dihydro-4H-pyrimido[5,4-b]indol-4-one. Single-crystal X-ray analysis and different spectroscopic techniques assured the assigned chemical structure of the title compound. In addition, molecular docking simulations and study were also executed for the title compound.

## 2. Results and Discussion

In frames of the development of a platform for biologically active compound libraries, design for actual biotargets, including platform testing on the example of the invention and preparation of candidate libraries for HBV treatment designed as inhibitors of viral penetration and assembly of viral core particles, we have carried out a pilot in vitro screening of a series of new compounds. At the first stage of the cytotoxicity pilot screening, all the potential inhibitors of viral infection were studied, of which about 1000 showed cytotoxicity over 50% at 10 µM. At the second stage of the screening from 27,297 substances with a cytotoxicity not exceeding 50%, 404 molecules were selected, which showed the ability to reduce the concentration of the HBV infection development marker HBeAg at the same concentration. It has been found that a quite promising molecular scaffold for the creation of a new potent drug for the treatment of hepatitis B is 3,5-dihydro-4H-pyrimido[5,4-b]indol-4-one (Figure 1).

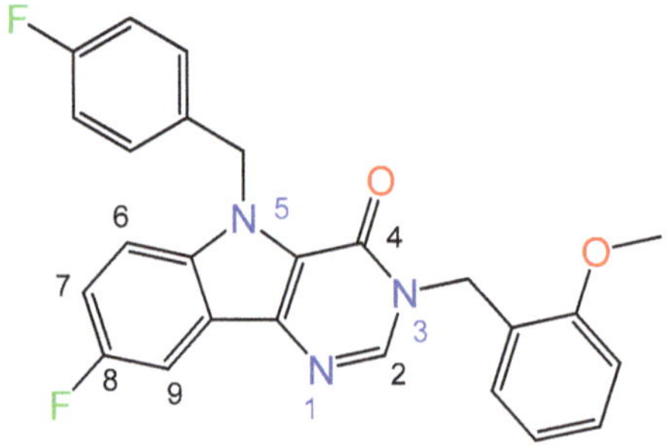

**Figure 1.** Chemical structure of title compound (3).

### 2.1. Synthesis

The synthesis of the title Compound (3) (Scheme 1) was carried out in several stages. The starting methyl 3-amino-5-fluoro-1-(4-fluorobenzyl)-1H-indole-2-carboxylate (1) was obtained according to the method described in Reference [23]. A further product (1) was converted to 8-fluoro-5-(4-fluorobenzyl)-3,5-dihydro-4H-pyrimido[5,4-b]indol-4-one (2) by boiling in formamide, and subsequent alkylation in DMF in the presence of sodium hydride led to the title Compound (3).

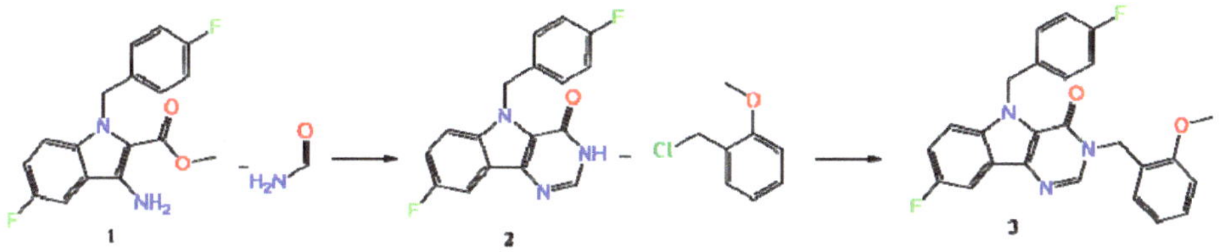

**Scheme 1.** Synthesis of the 8-fluoro-5-(4-fluorobenzyl)-3-(2-methoxybenzyl)-3,5-dihydro-4H-pyrimido [5,4-b]indol-4-one (**3**).

## 2.2. Crystal and Molecular Structure Analysis

The structure of Compound (**3**) was studied using different spectral methods (see Materials and Methods and Supporting information) and finally confirmed using XRD analysis.

The crystallographic information and refinement data are presented in Materials and Methods. Table S1 illustrates the bond angles and bond lengths (See Supporting information).

The unit cell asymmetric part of the studied crystal contained one molecule of the title Compound (**3**) (Figure 2). All atoms of the tricyclic fragment lay in the plane within 0.01 Å. The presence of vicinal substituents in the pyrimidinone cycle resulted in some nonequivalence of the $Csp^2$–N bonds: the C2-N1 bond (1.400(6) Å) was longer as compared to the C1-N1 bond (1.369(6) Å) and the mean value for the $Csp^2$–N bond (1.355 Å). The ortho-methoxy-phenyl fragment of the substituent at N1 atom was orthogonal to the bicyclic fragment (the C1-N1-C18-C19 torsion angle was 93.6 (4)°) and it was turned with respect to the N1-C18 bond (the N1-C18-C19-C20 torsion angle was −29.1(6)°). Such a position of this group resulted in the appearance of the H18B···O1 2.43 Å (the van der Waals radii sum was 2.46 Å) and H18B···O2 2.32 Å (2.46 Å) shortened intramolecular contacts, which cannot be considered as hydrogen bonds owing to the very small value of the C-H···O angles (97° and 102° respectively). The formation of the C20-H20···C1(π) intramolecular hydrogen bond (Table 3) was presumed to be stabilizing for such a position of this fragment. The methoxy group was slightly non-coplanar to the aromatic cycle (the C25-O2-C24-C23 torsion angle was 8.7(7)°) despite the essential repulsion between hydrogen atoms of the methyl group and cyclic atoms (the shortened intramolecular contacts were H23···H25A 2.32 Å (the van der Waals radii sum [24] was 2.34 Å), H25A···C23 2.78 Å (2.87 Å)). The para-fluorophenyl fragment of the substituent at N3 atom was located in the ac-conformation relatively to the C3-N3 endocyclic bond (the C3-N3-C11-C12 torsion angle was −78.3(5)°) and it was turned significantly with respect to the N3-C11 bond (the N3-C11-C12-C17 torsion angle was −50.7(5)°).

Most drugs are used in the solid-state, therefore special attention should be paid to the intermolecular interactions and features of the crystal structure.

In the crystal phase, molecules (**3**) formed centrosymmetric dimers (Figure 3) due to the C1-H1 ... F1 and C10-H10 ... N2 intermolecular hydrogen bonds (Table 1). These dimers were bound by stacking interactions (C ... C 3.29 Å and shift 4.99 Å, angle between planes 0°) and weak C25-H25A ... C21(π) intermolecular hydrogen bonds (Table 1) forming columns (Figure 4) along the [0 1 0] crystallographic direction. The molecules of the neighboring columns were bound by the C23-H23 ... C9(π) and C14-H14 ... O1 intermolecular hydrogen bonds (Table 1). Moreover, the H25 ... H14 intermolecular short contact (2.22 Å (2.32 Å)) has been revealed in the crystal.

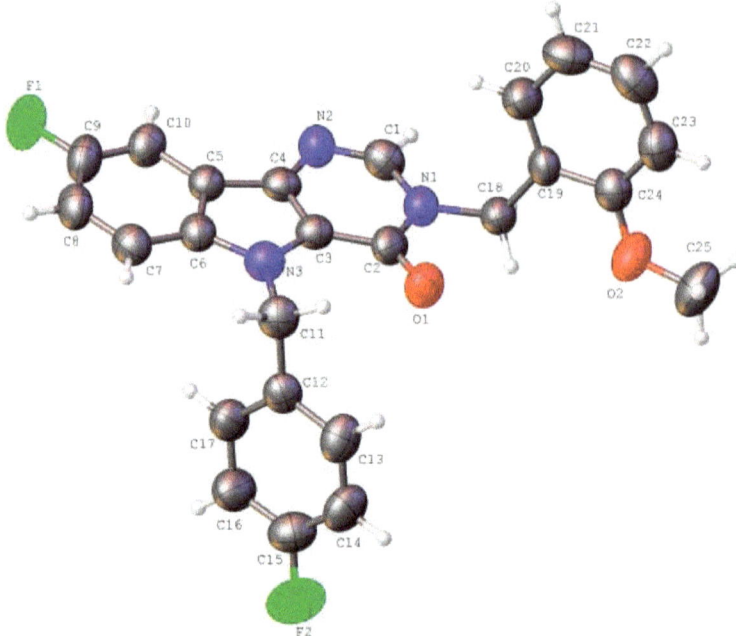

**Figure 2.** The molecular structure of the title Compound (3), showing the atom labeling. Displacement ellipsoids are drawn at the 50% probability level.

**Table 1.** Hydrogen bond characteristics (Å, °) in the crystal of Compound (3).

| D—H···A | D—H | H···A | D···A | D—H···A |
|---|---|---|---|---|
| C1—H1···F1 [i] | 0.93 | 2.60 | 3.433(5) | 149.0 |
| C10—H10···N2 [i] | 0.93 | 2.53 | 3.417(6) | 159.8 |
| C14—H14···O1 [ii] | 0.93 | 2.56 | 3.346(5) | 143.1 |
| C20—H20···C1 | 0.93 | 2.79 | 3.381(6) | 122.5 |
| C23—H23···C9 [iii] | 0.93 | 2.87 | 3.668(6) | 144.3 |
| C25-H25A ... C21 [iiii] | 0.93 | 2.74 | 3.572(3) | 144.9 |

Symmetry codes: (i) −x + 1, −y + 1, −z + 1; (ii) −x, −y + 2, −z + 1; (iii) x−1/2, −y + 3/2, z−1/2; (iiii) x, 1 + y, z.

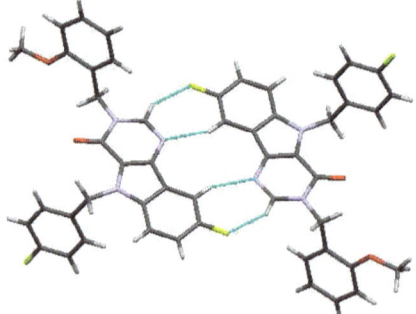

**Figure 3.** Centrosymmetric dimer in the crystal of titled Compound (3).

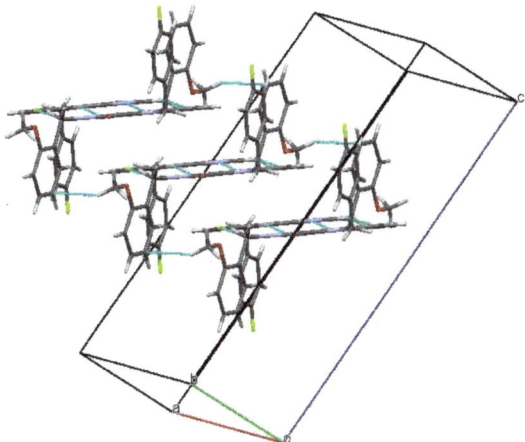

**Figure 4.** Columns of the dimers in the crystal phase.

*2.3. Hirshfeld Surface Analysis*

The peculiarities of the crystal packing were caused by the capability of the molecule to form certain types of intermolecular interactions. On the other hand, the biological activity depends on the formation of intermolecular interactions between the target molecule and the corresponding receptor. Thus the study of intermolecular interactions is a very important task.

One of the newest methods for intermolecular interactions analysis in the crystal phase is the study of Hirshfeld surfaces and 2D fingerprint plots generated using the CrystalExplorer program [25]. The Hirshfeld surfaces and their associated 2D fingerprint plots for the title Compound (**3**) were calculated taking the data of the single-crystal X-ray crystal structure as input. The Hirshfeld surface emerged from an attempt to define the space occupied by a molecule in a crystal for the purpose of partitioning the crystal electron density into molecular fragments [26]. It provided a 3D picture of the close contacts in the crystal, and these contacts can be summarized in a fingerprint plot [27]. The areas of the short contacts are shown by a red color on the Hirshfeld surfaces, while the long distances can be detected as the blue areas (Figure 5). The intermolecular interactions outward the H···H/C···H/F···H/O···H/N···H/C···C/C···N/C···F/C···O/N···F contacts, as well as the overall fingerprint region of the title molecule, are displayed in Figure 5. With this analysis, the division of contributions is possible for different interactions including H···H, N···H, C···H, and O···H, which commonly overlap in the full fingerprint plots. Figure 6 shows the $d_{norm}$ surface of Compound (**3**), the contribution from the C ... H/H ... C contacts, corresponding to the C—H ... π interaction, is represented by a pair of sharp spikes (29.0%). The absence of long sharp spikes indicates the absence of strong hydrogen bonds in the crystal of the title Compound (**3**) (Figure 6).

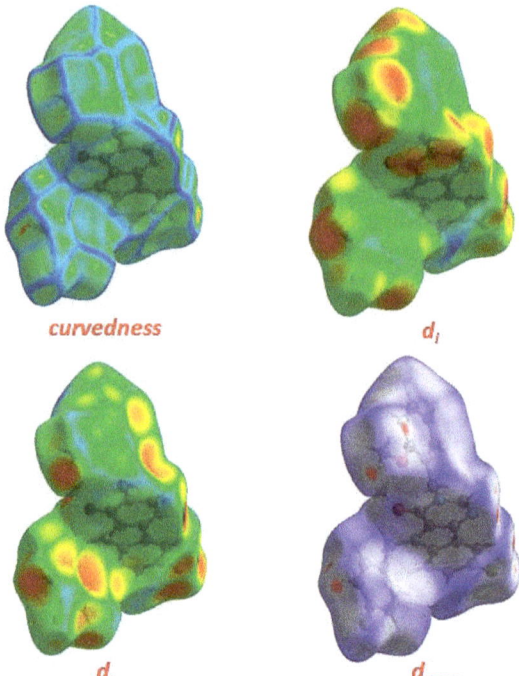

**Figure 5.** Hirshfeld surfaces for $d_{norm}$, $d_i$, and $d_e$ and curvedness for Compound (3).

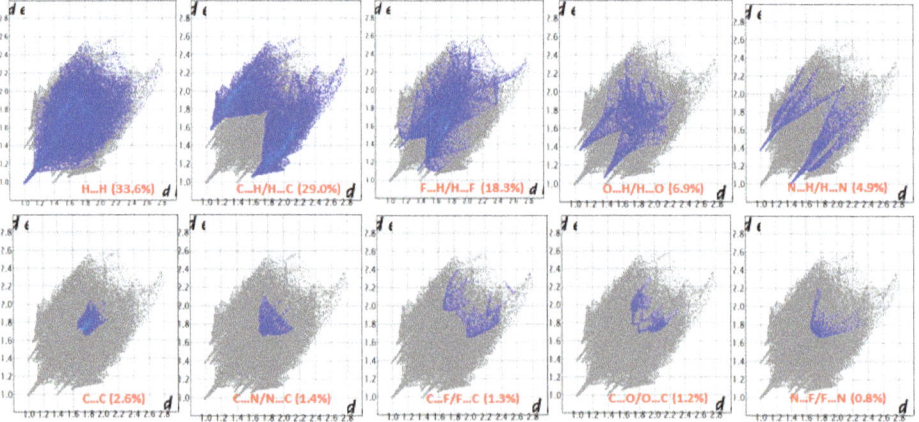

**Figure 6.** Two-dimension fingerprint plots showing various intermolecular contacts in Compound (3).

*2.4. Quantum Chemistry Calculations of Geometry and Electronic Structure*

Using computational methods to study properties of potentially biologically active molecules allows us to predict its behavior in the real environment. At the first stage, we have performed the optimization of the title molecule in a vacuum (using M062x DFT functional with cc-pVDZ basis set to estimate the influence of polar environment and crystal packing effects. Essentially we are interested in torsion angles values since they are closely related with effects of embedding of the ligand into the protein molecule.

As one can see, the calculated bond lengths were rather close to those obtained in the crystal phase (Table 2). However, the results for torsion angles, which correspond to rotation around single bonds, demonstrated significant differences between the X-ray and calculated values. It is obvious that such differences were connected with the packing effects.

**Table 2.** Geometrical parameters of the title compound (bond lengths in Å, torsion angles presented in degrees, numeration of atoms according to Figure 2).

| Bond/Torsion Angle | Experimental Value (X-ray Study) | Calculated Value, (Vacuum) |
|---|---|---|
| F1–C9 | 1.358(5) | 1.345 |
| F2–C15 | 1.366(5) | 1.342 |
| O1–C2 | 1.221(5) | 1.227 |
| O2–C24 | 1.362(5) | 1.354 |
| O2–C25 | 1.425(5) | 1.411 |
| N1–C1 | 1.369(5) | 1.377 |
| N1–C2 | 1.400(5) | 1.404 |
| N1–C18 | 1.460(4) | 1.472 |
| N2–C1 | 1.289(5) | 1.292 |
| N2–C4 | 1.375(5) | 1.371 |
| N3–C3 | 1.385(5) | 1.377 |
| N3–C6 | 1.395(5) | 1.381 |
| N3–C11 | 1.449(5) | 1.459 |
| C1–N1–C12–C19 | 93.6 | 58.5 |
| N1–C18–C19–C20 | −29.1 | −116.1 |
| C25–O2–C24–C23 | 8.7 | 1.7 |
| C3–N3–C11–C12 | −78.3 | −94.6 |
| N3–C11–C12–C17 | −50.7 | −106 |

In addition to the above mentioned geometric configuration (Conf. A), there were several conformations that are important in the description of the biological activity. They were generated by rotating around single bonds N3-C11, N1–C18 and C18–C19 (see Figure 2). Two additional conformations obtained in ab initio calculations are presented in Figure 7. The energy differences between the above-mentioned conformers were rather small—0.25 kcal/mol. Some comparisons of the ab initio results with the docking results will be presented in the next section.

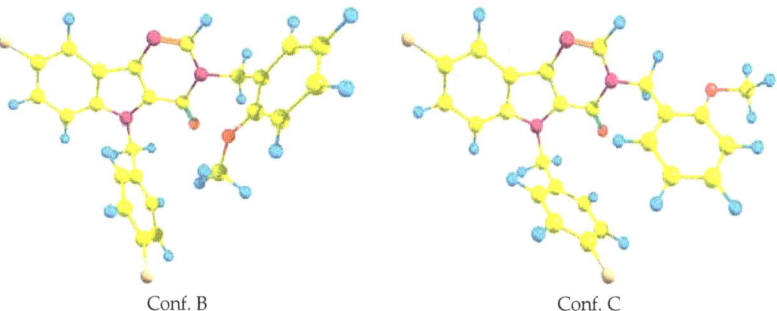

Conf. B      Conf. C

**Figure 7.** Two additional conformations of the title system (**3**).

In the structure-activity investigation, it is useful to describe the electronic structure of the molecule. The localization of the frontier molecular orbitals (FMOs) determines the reactivity and the ability to form intermolecular complexes. Therefore we present the orbital pictures for HOMO (highest occupied molecular orbital) and LUMO (lowest unoccupied molecular orbital) in Figure 8. Corresponding orbital energies were E(HOMO) = −6.8355, E(LUMO) = −0.5129 eV. The value of E(HOMO) and a rather large value of the HOMO–LUMO gap characterize the chemical system as relatively hard with electron donating properties.

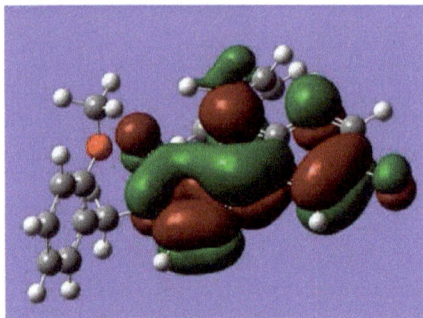 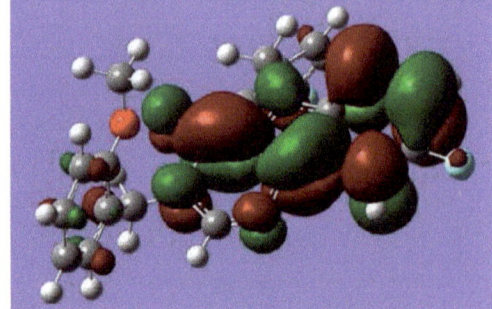

**Figure 8.** Graphical representation of frontier molecular orbitals (FMOs) of compound (3); left picture corresponds to highest occupied molecular orbital (HOMO), right—lowest unoccupied molecular orbital (LUMO).

According to our calculations, the HOMO and LUMO π-orbitals were localized at pyrimido [5,4-b]indol fragment of the molecule.

Good reproducibility of experimental data by quantum chemistry calculations indicates the possibility of using molecular modeling methods to estimate the potential interaction of the target molecule with receptors before conducting experimental tests on the biological activity.

*2.5. Molecular Docking Simulations*

The title molecule (3) has been tested as a system which interacts with the capsid of HBV (Hepatitis-B virus). Nowadays, the main approach to anti-HBV is connected with the development of new agents which are highly potent inhibitors of HBV replications [28]. Among the prospective agents which strongly interact with corresponding key proteins are derivatives of heteroaryldihydropyrimidine (HAP) [28–30]. Investigations of HAP (among them the most important chemical compounds—BAY41-4109, BAY39-5493, NVR-010-001-E2 [31–33]) demonstrated effective interactions with the corresponding HBV capsid and newly synthesized core protein. After interaction, the core protein could not assembly properly.

The compound we investigated in the present article can be treated as a structural analog of HAP. It is why we performed in silico modeling of interaction with corresponding core proteins. Fortunately for the different key ligand-proteins complexes (5T2P, 5WRE, 5GMZ, 5E0I), the X-ray crystallographic structures are known. In the present work, the 3D crystal structure of core proteins HBV bound to reference core molecules was retrieved from Reference [34].

The pharmacophore model and docking procedure have been performed by using Ligandscout 4.3. The software was provided by Reference [35]. All the above-mentioned structures of proteins contained six chains (designated as A, B, C, D, E, and F). At the first stage of our calculations, the residual mean square deviation (RMSD) between the docking-generated poses for the reference molecule and those obtained from the crystal structure X-ray investigations were calculated. As it was obtained the minimal values of RMSD, calculated for all the above-mentioned proteins, corresponded to the D-chain, where RMSD < 1 Å. The active site was found by using LigandScout option for pharmacophore generation. After that, the title molecule was docked at the corresponding site. The obtained data for score functions were collected in Table 3. In the table, the data corresponding to the reference core molecules is also presented. It can be seen that the title molecule interacted effectively with the protein. At the same time, the interaction parameters were comparable and even better than the corresponding values obtained for the reference molecules.

Table 3. Parameters of the binding affinity of the title molecule (3) for a site on the D chains.

| PDB | Molecule | Affinity (kcal/mol) | Binding Affinity Score |
|---|---|---|---|
| 5E0I | NVR10-001E2 | −21.6 | −28.6 |
| | molecule (3) | −17.9 | −33.5 |
| 5GMZ | 4-methyl heteroaryldihydropyrimidine | −16.18 | −25.95 |
| | molecule (3) | −16.30 | −31.02 |
| 5WRE | HAP-R01 | −21.46 | −38.45 |
| | molecule (3) | −15.16 | −23.39 |
| 5T2P | SBA-R01 | −15.81 | −26.70 |
| | molecule (3) | −17.14 | −44.42 |

As an example, the position receptor-ligands complex (5E0I) can be seen in Figure 9. The 2D representation of the structural formula of the ligands demonstrated its correspondence to the generated pharmacophore. Here, the hydrogen bond acceptor is designated by the red dotted line, while the hydrophobic interaction is designated by the yellow line. For the detailed descriptions of the pharmacophore designation and generation, see the LigandScout manual [35].

The geometric structure of the ligand in this complex for case A was qualitatively close to conformation C while for case B—to the conformation B (see Figure 7). Comparison of most important torsion angles is presented in Table 4. Here, the numeration of atoms is in accordance with Figure 2.

Table 4. Torsion angles of ligands obtained by ab initio and docking (5E0I protein) calculations.

| Torsion Angle | Docking Case A | ab Initio Conf. C | Docking Case B | ab Initio Conf. B |
|---|---|---|---|---|
| N3–C11–C12–C17 | −3.6 | 53.6 | 30.9 | 19 |
| N1–C18–C19–C24 | −176.1 | −170 | 63.2 | 60 |

As one can see, the torsion angle N3–C11–C12–C17 underwent significant changes when the ligand was incorporated into the protein.

The simulation study of the 3D binding of the title molecule with corresponding proteins demonstrated effective interaction and hence can be treated as a potent inhibitor of HBV replications.

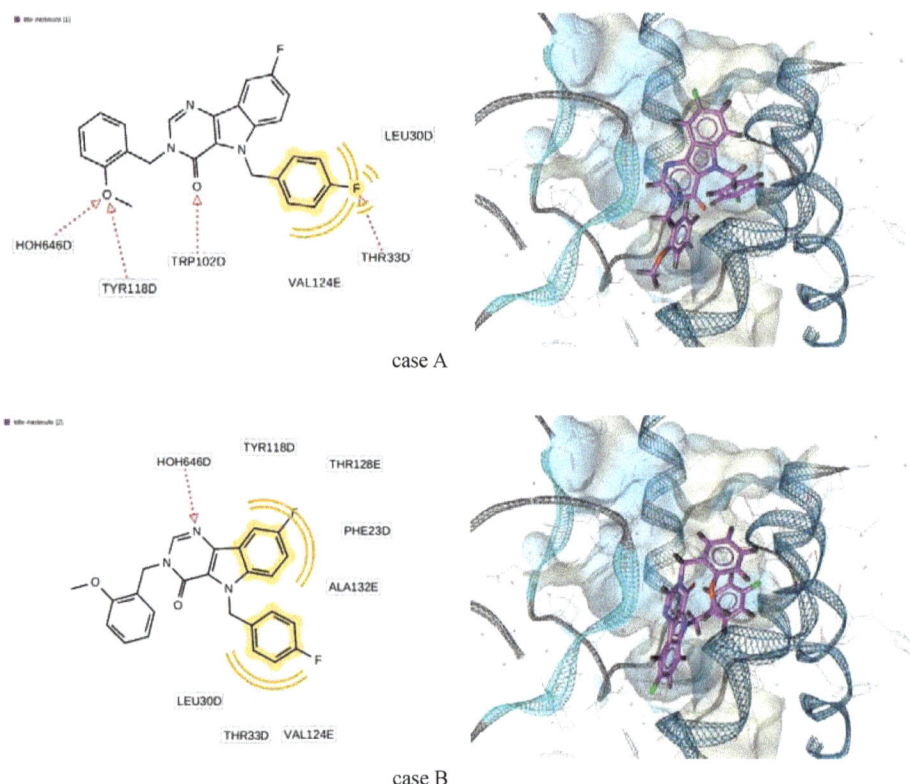

**Figure 9.** Generated docking pose for complex "title molecule (3)—5E0I".

## 2.6. Anti-Hepatitis B Virus (HBV) Activity

The final stage of our investigation was the experimental study of the biological activity of molecule (3). The human hepatoma cell line HepAD38 was chosen as the source of infectious viral particles to infect the HepG2/NTCP cell line, carrying the stable integrated HBV virus gene under the control of a tetracycline-regulated promoter, and secreting viral particles into the culture medium in the absence of tetracycline. HBV preparation was obtained using the HepAD38 line according to the following protocol: HepAD38 cells were passaged in a DMEM medium containing 10% fetal calf serum, penicillin/streptomycin and essential amino acids. The culture medium was taken once every 2 days, clarified by centrifugation (200× $g$, 15 min) and stored at 4 °C for no longer than 7 days. Next, dry PEG 8000 was added to the culture media to a final concentration of 7.5% and incubated at 4 °C on a rotary platform overnight. The viral precipitate was separated by centrifugation (2000× $g$, 30 min) and the precipitate was suspended in 1/100 of the initial volume in OPTI-MEM medium. Thus obtained viral preparation was aliquoted and stored at −80 °C.

Infection was carried out as follows: The HepG2-NTCP cell suspension was distributed to 96-well plates at 2000 cells per well. After the cells were attached (on the same or the next day), the initial solution was removed by aspiration, and 50 µL of a solution of test compounds dissolved in OPTI-MEM medium (with a final DMSO concentration of 2%) was added to each well or OPTI-MEM with 2% DMSO (in the wells of the positive and negative controls of the infection) and 50 µL of the HBV preparation diluted in OPTI-MEM with 2% DMSO (except negative infection control). After incubation for 24 h in a humidified atmosphere containing 5% $CO_2$, the HBV medium was removed by aspiration, and 200 µL of DMEM culture medium containing the corresponding test compounds in 10 mkM concentration was added to the cultures. The cells were additionally incubated for 6 days at 37 °C in a humidified atmosphere containing 5% carbon dioxide. Next, cell supernatants (50 µL) were analyzed for viral antigen content using a commercial HBeAg ELISA 4.0 kit (Creative Diagnostics, catalog number DEIA003) according to the kit manufacturer's protocol and the optical density of each analyzed well was measured at a wavelength of 450 nm using a plate densitometer.

The title molecule (3) in an experimental in vitro model of hepatitis B virus infection with the usage of human hepatoma line HepG2, was stably transfected with the NTCP gene to impart the ability to be infected with HBV and maintain a full virus replication cycle demonstrated 83 % inhibition (for 10 mkM concentration).

## 3. Materials and Methods

*3.1. Chemistry*

3.1.1. General Information

All chemicals were obtained from Sigma-Aldrich or Merck. Detailed procedure of starting methyl 3-amino-5-fluoro-1-(4-fluorobenzyl)-1H-indole-2-carboxylate (1) synthesis has been reported [11]. NMR spectra were registered on a Bruker DPX 300 spectrometer (Bruker BioSpin AG, Industriestrasse 26, 8117 Fällanden, Switzerland) at room temperature (298 K) on a using DMSO-d6 as a solvent and processed using Bruker XWinNMR software (International Equipment trading Ltd., Mestrenova Version 9.0 Spain, Vernon Hills, IL 60061, USA). Liquid chromatography mass spectrometry was developed by means of chromatography with PHENOMENEX GEMINI NX C18 110 Å 4.61 × 150 mm column (0.05% TFA, gradient ACN/H$_2$O), UV-detector SHIMADZU SPD-10AD VP (registered absorption at 254 nm), ELSD (evaporative light scattering detector) SEDEX-75 (S.E.D.E.R.E., PARC VOLTA-BP 27 9, Rue Parmentier 94141 Alfortville Cedex FRANCE) and API-150EX mass-spectrometer (MDS Sciex, Toronto, ON, Canada). Elution started with 0.1 M solution of TFA in water and ended with 0.1 M solution of TFA in acetonitrile used a linear gradient at a flow rate of 0.15 mL/min and an analysis cycle time of 25 min. The FT-IR spectrum was registered in a KBr pellet with a Shimadzu IR Prestige-21 Fourier Transform Infrared (FTIR) Spectrophotometer (Shimadzu Corporation, Kyoto, Japan). The UV/Vis spectrum was registered in acetonitrile with an Agilent 8453 UV-visible Spectrophotometer. The melting point was registered with a Buchi M-560 (Buchi AG, Meierseggstrasse 40, 9230 Flawil, Switzerland). Elemental analysis was performed on EuroEA-3000 CHNS-O Analyzer (Euro Vector, Via Tortona 5-20144, Milan, Italy).

3.1.2. Synthesis and Crystallization of Compound (3)

To a solution of compound (2) (1.0 mmol, 311 mg) in DMF (5 mL), 1.1 mmol of NaH was added. After 0.5 h, 1-(chloromethyl) 2 methoxybenzene (1.1 mmol, 172 mg) was added dropwise to the mixture and stirred at 50 °C for 1 h. The mixture was heated for 1 h at 50 °C and then cooled down to room temperature. The solution was poured into water. The formed precipitate was collected by filtration, recrystallized from acetonitrile, filtered off, washed with acetonitrile and dried to give pure product (3). Yield 336 mg (78%), white crystallic powder, m.p. 154.1–154.6 °C; UV (MeOH) λmax (nm) 223, 270, 288, 298, 325, 335, 350 (Figure S1); IR (KBr): ν (cm$^{-1}$) 3052, 2943 2839 (C–H), 1669 (C=O), 1588, 1565, 1504, 1461, 1371, 1305, 1243, 1214, 1022, 859, 795, 754 (Figure S2); $^1$H NMR (300 MHz, DMSO-d6): δ 8.42 (s, 1H, H-2), 7.77 (dt, *J* = 8.6, 3.0 Hz, 2H), 7.39 (td, *J* = 9.2, 2.6 Hz, 1H), 7.34–7.20 (m, 3H), 7.12–7.00 (m, 4H), 6.90 (td, *J* = 7.5, 1.1 Hz, 1H), 5.91 (s, 2H, 5-CH$_2$), 5.24 (s, 2H, 3-CH$_2$), 3.84 (s, 3H, OCH$_3$); $^{13}$C NMR (75 MHz, DMSO-d6): δ 44.85 (3-CH$_2$), 46.64 (5-CH$_2$), 55.50 (OCH$_3$), 105.19 (d, *J* = 23.7 Hz, C-9), 110.95, 113.15 (d, *J* = 9.3 Hz), 115.37 (d, *J* = 21.4 Hz), 116.32 (d, *J* = 26.6 Hz), 120.34, 121.00 (d, *J* = 9.9 Hz), 122.12, 124.26, 128.84, 129.08, 129.19, 134.15 (d, *J* = 3.1 Hz), 135.91, 137.53 (d, *J* = 5.2 Hz), 145.37, 154.56, 156.98, 157.53 (d, *J* = 236.6 Hz, 8-CF), 161.42 (d, *J* = 243.5 Hz, 4'-CF); $^{19}$F NMR (377 MHz, DMSO-d6): δ

−121.77 (d, $J$ = 4.3 Hz), −115.14 (Figure S3); LC/MS $m/z$ (%): 432.5 [MH]$^+$ (99); found, %: C 69.93; H 4.40; N 9.77. $C_{25}H_{19}F_2N_3O_2$. Calculated, %: 69.60; H 4.44; N 9.74.

Further crystallization by slow evaporation of a solution in acetonitrile was carried out to provide single stick-like colorless crystals suitable for X-ray diffraction analysis (Figure 10).

**Figure 10.** Crystals of the title Compound (**3**).

*3.2. X-ray Diffraction Study*

The crystals of (**3**) ($C_{25}H_{19}F_2N_3O_2$) were monoclinic. At 293 K $a$ = 16.366(3) Å, $b$ = 6.0295(14) Å, $c$ = 21.358(4) Å, $\beta$ = 105.21(2)°, $V$ = 105.21(2) Å$^3$, $M_r$ = 431.43, $Z$ = 4, space group $P2_1/n$, $d_{calc}$ = 1.409 g/cm$^3$, $\mu(MoK_\alpha)$ = 0.103 mm$^{-1}$, $F(000)$ = 896. Intensities of 14,544 reflections (3571 independent, $R_{int}$ = 0.1315) were measured on the «Xcalibur-3» diffractometer (graphite monochromated MoK$\alpha$ radiation, CCD detector, $\omega$-scaning, 2$\theta$max = 50°).

The structure was solved by direct methods using the SHELXTL package [36]. The position of the hydrogen atoms was located from electron density difference maps and refined by the "riding" model with $U_{iso}$ = $nU_{eq}$ of the carrier atoms (n = 1.5 for methyl groups and n = 1.2 for other hydrogen atoms). Full-matrix least-squares refinement of the structures against $F^2$ in anisotropic approximation for non-hydrogen atoms using 3571 reflections was converged to: $wR_2$ = 0.184 ($R_1$ = 0.068 for 1558 reflections with $F > 4\sigma(F)$, S = 0.913).The final atomic coordinates and crystallographic data for molecule (**3**) have been deposited to with the Cambridge Crystallographic Data Centre, 12 Union Road, CB2 1EZ, UK (fax: +44-1223-336033; e-mail: deposit@ccdc.cam.ac.uk) and are available on request quoting the deposition number CCDC 1915665).

*3.3. Theoretical Calculations*

Crystal Explorer 17.5 was utilized to generate fingerprint plots and Hirshfeld surface map of the title compound (**3**) [25]. All quantum chemical calculations were carried out using Gaussian 09 [37].

*3.4. Docking Studies*

The pharmacophore model generation and docking were performed using Ligandscout 4.3 program [35].

## 4. Conclusions

The new biologically active compound (8-fluoro-5-(4-fluorobenzyl)-3-(2-methoxybenzyl)-3,5-dihydro-4H-pyrimido[5,4-b]indol-4-one) has been studied using both theoretical and experimental methods. The title compound was characterized by spectral methods and its molecular and crystal structure has been confirmed by X-ray diffraction study. Being very important to provide the interaction with receptors intermolecular interactions were studied thoroughly. Some approaches to quantum chemical modeling of possible interaction of the title molecule with receptors have been used. At he first stage, it was shown the reproducibility of experimental molecular geometry by the quantum chemical calculations. At the next stage, the interactions between the potential drug and receptors were modeled. Finally, the experimental study of biological activity has shown that a very potent drug for hepatitis B has been developed.

**Supplementary Materials:** The following are available online at http://www.mdpi.com/2073-4352/9/8/379/s1, Figure S1: UV/Vis spectrum of the title compound (3) (MeOH, M 0.0116 mmol/l, optical cell 1.0 cm), Figure S2: IR spectrum of the title compound (3) (KBr pellet), Figure S3: 19F NMR (376.72 MHz, DMSO-d6) spectrum of the title compound (3), Figure S4: 13C NMR (75.48 MHz, DMSO-d6) spectrum of the title compound (3), Figure S5: 1H NMR (300.16 MHz, DMSO-d6) spectrum of the title compound (3), Figure S6: LC/MS Data for Structural Determination of of the title compound (3), Table S1: Geometric parameters (Å, ○) of the title Compound (3).

**Author Contributions:** A.V.I. proposed the work and prepared the manuscript for publication, S.M.K. and D.V.K. synthesized and characterized the title molecule, T.L., I.V.K. and N.D.B. conducted the computational calculations, O.D.M. performed the screening of anti-HBV activity. All authors discussed the contents of the manuscript.

**Funding:** The work was supported by the Ministry of Science and Higher Education of the Russian Federation in frames of agreement on reimbursement of costs associated with the development of a platform for biologically active compound libraries and the design of actual biotargets, including platform testing on the example of invention and preparation of candidate libraries for HBV treatment designed as inhibitors of viral penetration and assembly of viral core particles (RFMEFI57917X0154).

**Acknowledgments:** We thank A.N. Nesmeyanov Institute of Organoelement Compounds of Russian Academy of Sciences (INEOS RAS) for the support of the Crystallography Laboratory. The authors are grateful to Konovalova I.S. (SSI "Institute for Single-crystals", Kharkiv) for assistance with the XRD analysis.

**Conflicts of Interest:** The authors declare no conflict of interest.

## References

1. World Health Organization (WHO). *Hepatitis B Fact Sheet N204*; WHO: Geneva, Switzerland, 2017.
2. Elguero, J.; Goya, P.; Jagerovic, N.; Silva, A.M.S. Pyrazoles as Drugs: Facts and Fantasies in Targets in Heterocyclic Systems. *Ital. Soc. Chem.* **2002**, *6*, 52.
3. Dewick, P.M. *Medicinal Natural Products: A Biosynthetic Approach*, 2nd ed.; Wiley: New York, NY, USA, 2002.
4. Deiters, A.; Martin, S.F. Synthesis of Oxygen- and Nitrogen-Containing Heterocycles by Ring-Closing Metathesis. *Chem. Rev.* **2004**, *104*, 2199. [CrossRef] [PubMed]
5. Agarwal, S.; Cämmerer, S.; Filali, S.; Fröhner, W.; Knöll, J.; Krahl, M.P.; Reddy, K.R.; Knölker, H.-J. Novel Routes to Pyrroles, Indoles and Carbazoles - Applications in Natural Product Synthesis. *Curr. Org. Chem.* **2005**, *9*, 1601. [CrossRef]
6. Gataullin, R.R. Synthesis of compounds containing a cycloalka[b]indole fragment. *Russ. J. Org. Chem.* **2009**, *45*, 321. [CrossRef]
7. Bolton, D.; Forbes, I.T.; Hayward, C.J.; Piper, D.C.; Thomas, D.R.; Thompson, M.; Upton, N. Synthesis and potential anxiolytic activity of 4-amino-pyrido [2, 3-b] indoles. *Bioorg. Med. Chem. Lett.* **1993**, *3*, 1941. [CrossRef]
8. Love, B.E. Synthesis of carbolines possessing antitumor activity. *Top. Heterocycl. Chem.* **2006**, *2*, 93.
9. Somei, U.; Basha, A. *Indole Alkaloids*; Hawood Academic Publishers. Amsterdam, The Netherlands, 1997.
10. Humphrey, G.R.; Kuethe, J.T. Practical methodologies for the synthesis of indoles. *Chem. Rev.* **2006**, *106*, 2875. [CrossRef]
11. Sundberg, R.J. *Indoles*, 1st ed.; Academic Press: London, UK, 1996.
12. Sundberg, R.J. *Comprehensive Heterocyclic Chemistry II*, 2nd ed.; Katritzky, A.R., Ress, C.W., Scriven, E.F.V., Bird, C.W., Eds.; Pergamon Press: Oxford, UK, 1996; Volume 2, p. 119.
13. Okamoto, A.; Tanaka, K.; Saito, I. Rational design of a DNA wire possessing an extremely high hole transport ability. *J. Am. Chem. Soc.* **2003**, *125*, 5066. [CrossRef]
14. Okamoto, A.; Tanaka, K.; Saito, I.J. DNA logic gates. *Am. Chem. Soc.* **2004**, *126*, 9458. [CrossRef]
15. Showalter, D.H.H.; Bridges, J.A.; Zhou, H.; Sercel, D.A.; McMichael, A.; Fry, D.W.J. Tyrosine Kinase Inhibitors. 16. 6,5,6-Tricyclic Benzothieno[3,2-d]pyrimidines and Pyrimido[5,4-b]- and -[4,5-b]indoles as Potent Inhibitors of the Epidermal Growth Factor Receptor Tyrosine Kinase. *Med. Chem.* **1999**, *42*, 5464. [CrossRef]
16. Bundy, G.L.; Banitt, L.S.; Dobrowoski, P.L.; Palmer, J.R.; Schwartz, T.M.; Zimmermann, D.C.; Lipton, M.F.; Mauragis, M.A.; Veley, M.F.; Appell, R.B.; et al. Synthesis of 2, 4-Di-1-pyrrolidinyl-9H-pyrimido [4, 5-b] indoles, including antiasthma clinical candidate PNU-142731A. *Org. Process Res. Dev.* **2001**, *5*, 144. [CrossRef]

17. Barnes, P.J.; Chung, K.F.; Page, C.P. Inflammatory mediators of asthma: An update. *Pharmacol. Rev.* **1998**, *50*, 515. [PubMed]
18. Howarth, P.H. The airway inflammatory response in allergic asthma and its relationship to clinical disease. *Allergy* **1995**, *50*, 13. [CrossRef] [PubMed]
19. Muller, C.E.; Hide, I.; Daly, J.; Rothenhausler, K.; Eger, K. 7-Deaza-2-phenyladenines: structure-activity relationships of potent A1 selective adenosine receptor antagonists. *J. Med. Chem.* **1990**, *33*, 2822. [CrossRef] [PubMed]
20. Darrow, J.W.; Maynard, G.D.; Horvath, R.F. Substituted 9h-pyridino[2,3-b]indole and 9h-pyrimidino[4,5-b]indole Derivatives: Selective Neuropeptide y Receptor Ligands. International Patent Application Publication No. WO 9951598, 14 October 1999.
21. Ivachtchenko, A.V.; Ivachtchenko, A.A.; Savchuk, N.P.; Rogovoj, B.; Bychko, V.V. Hepatitis B Virus (HBV) Inhibitor. International Patent Application Publication No. WO 2019017814, 24 January 2019.
22. Ivachtchenko, A.V.; Ivachtchenko, A.A.; Savchuk, N.P.; Rogovoj, B.; Bychko, V.V.; Khvat, A. Hepatitis B virus (HBV) Penetration Inhibitor and Pharmaceutical Composition for Hepatitis Treatment. International Patent Application Publication No. RU 2662161, 24 July 2018.
23. Barker, A.J.; Kettle, J.G.; Faull, A.W. Preparation of Substituted Indoles for Treatment of A Disease or Condition Mediated by Monocyte Chemoattractant Protein-1 (MCP-1). International Patent Application Publication No. WO 9907351 A2, 18 February 1999.
24. Zefirov, Y.V.; Zorky, P.M. New applications of van der Waals radii in chemistry. *Russ. Chem. Rev.* **1995**, *64*, 415–428. [CrossRef]
25. Turner, M.J.; McKinnon, J.J.; Wolff, S.K.; Grimwood, D.J.; Spackman, P.R.; Jayatilaka, D.; Spackman, M.A. *Crystal Explorer 17*; The University of Western Australia: Perth, Australia, 2017.
26. Spackman, M.A.; Byrom, P.G. A novel definition of a molecule in a crystal. *Chem. Phys. Lett.* **1997**, *267*, 215–220. [CrossRef]
27. Spackman, M.A.; Jayatilaka, D. Hirshfeld surface analysis. *Cryst. Eng. Comm.* **2009**, *11*, 19–32. [CrossRef]
28. Endres, D.; Miyahara, M.; Moisant, P.; Zlotnick, A. A reaction landscape identifies the intermediates critical for self-assembly of virus capsids and other polyhedral structures. *Protein Sci.* **2005**, *14*, 1518–1525. [CrossRef]
29. Stray, S.J.; Zlotnick, A. BAY 41-4109 has multiple effects on Hepatitis B virus capsid assembly. *J. Mol. Recognit.* **2006**, *19*, 542–548. [CrossRef]
30. Choi, I.G.; Yu, Y.G. Interaction and assembly of HBV structural proteins: Novel target sites of anti-HBV agents. *Infect. Disord. Drug Targets* **2007**, *7*, 251–256. [CrossRef]
31. Firdayani, A.; Arsianti, C.; Yanuar, A. Molecular Docking and Dynamic Simulation Benzoylated Emodin into HBV Core Protein. *J. Young Pharm.* **2018**, *10*, S20–S24. [CrossRef]
32. Wu, G.; Liu, B.; Zhang, Y.; Li, J.; Arzumanyan, A.; Clayton, M.M.; Schinazi, R.F.; Wang, Z.; Goldmann, S.; Ren, Q.; et al. Preclinical characterization of GLS4, an inhibitor of hepatitis B virus core particle assembly. *Antimicrob. Agents Chemother.* **2013**, *57*, 5344–5354. [CrossRef] [PubMed]
33. Wang, X.Y.; Wei, Z.M.; Wu, G.Y.; Wang, J.H.; Zhang, Y.J.; Li, J.; Zhang, H.H.; Xie, X.W.; Wang, X.; Wang, Z.H.; et al. In vitro inhibition of HBV replication by a novel compound, GLS4, and its efficacy against adefovir-dipivoxil-resistant HBV mutations. *Antivir. Ther.* **2012**, *17*, 793–803. [CrossRef] [PubMed]
34. Berman, H.M.; Westbrook, J.; Feng, Z.; Gilliland, G.; Bhat, T.N.; Weissig, H.; Shindyalov, I.N.; Bourne, P.E. The Protein Data Bank Nucleic Acids Research. 2000, Volume 28, pp. 235–242. Available online: http://www.rcsb.org/ (accessed on 24 July 2019).
35. Wolber, G.; Langer, T. LigandScout: 3-D Pharmacophores Derived from Protein-Bound Ligands and Their Use as Virtual Screening Filters. *J. Chem. Inf. Model.* **2005**, *45*, 160–169. Available online: http://www.inteligand.com/ligandscout/ (accessed on 24 July 2019). [CrossRef] [PubMed]
36. Sheldrick, G.M. SHELXT—Integrated space-group and crystal-structure determination. *Acta Cryst.* **2015**, *A71*, 3–8. [CrossRef] [PubMed]
37. Frisch, M.J.; Trucks, G.W.; Schlegel, H.B.; Scuseria, G.E.; Robb, M.A.; Cheeseman, J.R.; Scalmani, G.; Barone, V.; Mennucci, B.; Petersson, G.A.; et al. *Gaussian-09; Revision A.02*, Gaussian, Inc.: Wallingford, CT, USA, 2009.

© 2019 by the authors. Licensee MDPI, Basel, Switzerland. This article is an open access article distributed under the terms and conditions of the Creative Commons Attribution (CC BY) license (http://creativecommons.org/licenses/by/4.0/).

Article

# (2*E*)-2-[1-(1,3-Benzodioxol-5-yl)-3-(1*H*-imidazol-1-yl) propylidene]-*N*-(2-chlorophenyl)hydrazine carboxamide: Synthesis, X-ray Structure, Hirshfeld Surface Analysis, DFT Calculations, Molecular Docking and Antifungal Profile

Reem I. Al-Wabli [1,*], Alwah R. Al-Ghamdi [1], Suchindra Amma Vijayakumar Aswathy [2], Hazem A. Ghabbour [3], Mohamed H. Al-Agamy [4,5], Issac Hubert Joe [2] and Mohamed I. Attia [1,6,*]

[1] Department of Pharmaceutical Chemistry, College of Pharmacy, King Saud University, P.O. Box 2457, Riyadh 11451, Saudi Arabia; toto24se@hotmail.com
[2] Center for Molecular Biophysics Research, Mar Ivanios College, Nalanchira, Thiruvanthapuram 695015, India; aswathyrs86@gmail.com (S.A.V.A.); hubertjoe@gmail.com (I.H.J.)
[3] Department of Medicinal Chemistry, Faculty of Pharmacy, Mansoura University, Mansoura 35516, Egypt; ghabbourh@yahoo.com
[4] Department of Pharmaceutics, College of Pharmacy, King Saud University, P.O. Box 2457, Riyadh 11451, Saudi Arabia; malagamy@KSU.EDU.SA
[5] Microbiology and Immunology Department, Faculty of Pharmacy, Al-Azhar University, Cairo 11884, Egypt
[6] Medicinal and Pharmaceutical Chemistry Department, Pharmaceutical and Drug Industries Research Division, National Research Centre (ID: 60014618), El Bohooth Street, Dokki, Giza 12622, Egypt
* Correspondence: ralwabli@KSU.EDU.SA (R.I.A.-W.); mattia@ksu.edu.sa (M.I.A.); Tel.: +966-11-805-2620 (R.I.A.-W.); +966-146-77337 (M.I.A.)

Received: 3 January 2019; Accepted: 29 January 2019; Published: 4 February 2019

**Abstract:** Life-threatening fungal infections accounts for a major global health burden especially for individuals suffering from cancer, acquired immune deficiency syndrome (AIDS), or autoimmune diseases. (2*E*)-2-[1-(1,3-Benzodioxol-5-yl)-3-(1*H*-imidazol-1-yl)propylidene]-*N*-(2-chlorophenyl) hydrazinecarboxamide has been synthesized and characterized using various spectroscopic tools to be evaluated as a new antifungal agent. The (*E*)-configuration of the imine moiety of the title molecule has been unequivocally identified with the aid of single crystal X-ray analysis. The molecular structure of compound **4** was crystallized in the monoclinic, $P2_1/c$, $a = 8.7780~(6)$ Å, $b = 20.5417~(15)$ Å, $c = 11.0793~(9)$ Å, $\beta = 100.774~(2)°$, $V = 1962.5~(3)$ Å$^3$, and $Z = 4$. Density functional theory computations have thoroughly explored the electronic characteristics of the title molecule. Moreover, molecular docking studies and Hirshfeld surface analysis were also executed on the title compound **4**. The in vitro antifungal potential of the target compound was examined against four different fungal strains.

**Keywords:** Crystal structure; Imidazole; Benzodioxole; Semicarbazone; DFT

## 1. Introduction

The dramatic increase in the morbidity due to life-threatening fungal infections accounts for a major health burden worldwide, particularly for those individuals suffering from cancer, AIDS, or autoimmune diseases [1–3]. The presence of few drug targets in the eukaryotic fungal cells that are not shared with human cells limits the availability of selective antifungal agents, as compared with the anti-bacterials [4]. The emergence of resistance to the available antifungal agents as well as their serious side effects led to the necessity for the search for new alternative antifungal therapies [5].

Imidazole and/or triazole moieties constitute the core pharmacophore part of most azoles, which are used as a first-line treatment for various fungal infections [6]. Azoles inhibit competitively the sterol biosynthesis via inhibition of fungal lanosterol 14α-demethylase (CYP51) leading to a hindering of the normal fungal growth [7]. The presence of two carbons linking the azole pharmacophore part and an aromatic residue is a common feature in the available antifungal azoles. Whereas, the reported antifungal candidates bearing a propyl bridige between the aromatic moiety and the azole fragment are limited [8–10].

Semicarbazones are formally derived from the condensation of certain semicarbazide with the appropriate aldehydes or ketones. While semicarbazones have received less attention than their respective thiosemicarbazones, they became versatile ligands and drawn the attention of a number of researchers in the scientific community during the last fifteen years. Semicarbazones have the ability to coordinate with various metal ions and form complexes due to the existence of oxygen and nitrogen donor atoms in their core structure. In addition, they mostly occur in the keto form in the solid state, while they exhibit a keto-enol tautomerism in the solution state [11]. Semicarbazones displayed significant antibacterial, anticonvulsant, antioxidant, and antifungal activities and their structures can be further elaborated to prepare various bioactive heterocyclic compounds [12,13]. Moreover, metal complexes of semicarbazones exhibited a broad spectrum bioactivities like the activities toward smallpox, protozoa, influenza, trypanosomiasis, certain kinds of tumors, malaria, and fungi [14].

In addition, literature search exposed that benzodioxole fragment occurs in a large number of biologically active molecules including antimicrobials [15–17].

In a continuation of our efforts to obtain new potent antifungal compounds [18–22], we were inspired to design and synthesize (2E)-2-[1-(1,3-benzodioxol-5-yl)-3-(1H-imidazol-1-yl) propylidene]-N-(2-chlorophenyl)hydrazinecarboxamide to be examined as a new antifungal agent. The title compound comprises three pharmacophore parts, namely imidazole, semicarbazone, and benzodioxole with a three-carbon bridge separating benzodioxole and imidazole fragments with the aim to get new potent drug-like antifungal hybrid. Single crystal X-ray analysis and different spectroscopic techniques assured the assigned chemical structure of the title compound as well as the (E)-configuration of its imine functionality. In addition, the electronic characters and molecular geometry of the title molecule were explored with the aid of density functional theory (DFT) computations. Molecular docking simulations and Hirshfeld surface analysis were also executed for the title compound 4. The in vitro antifungal profile of the target compound was examined against four different fungal strains.

## 2. Experimental

### 2.1. General

Gallenkamp melting point apparatus was harnessed to measure melting point of compound 4 and it is uncorrected. Nuclear magnetic resonance (NMR) measurements were executed in DMSO-$d_6$ with the aid of Bruker NMR spectrometer at 500 MHz for $^1$H and 125.76 MHz for $^{13}$C at the Research Center, College of Pharmacy, King Saud University, Saudi Arabia. δ-values (ppm) were used to represent chemical shifts in relation to tetramethylsilane (TMS) as an internal standard. Mass spectrum of compound 4 was recorded using Agilent Quadrupole 6120 LC/MS (Agilent Technologies, Palo Alto, CA, USA) implemented with ESI (Electrospray ionization) source. Silica gel precoated aluminum thin layer chromatography plates with fluorescent indicator at 254 nm were secured from Merck (Darmstadt, Germany) and illumination with UV light source (254 nm) accomplished visualization.

## 2.2. Synthesis

Synthesis of (2E)-2-[1-(1,3-benzodioxol-5-yl)-3-(1H-imidazol-1-yl)propylidene]-N-(2-chloro phenyl)hydrazinecarboxamide (4)

Few drops of glacial acetic acid were added to a stirred reaction mixture containing ketone 3 (0.24 g, 10 mmol) and N-(2-chlorophenyl)hydrazinecarboxamide (1.86 g, 10 mmol) [23], in absolute ethyl alcohol (15 mL) and stirring was continued atambient temperature for 18 h. Ethyl alcohol was removed under vacuum and the residue was purified by re-crystallization from ethyl alcohol to afford the corresponding semicarbazones 4 as colorless crystals suitable for X-ray analysis. Yield 0.51 g (51%); white powder m.p. 163–165 °C; IR (KBr): $\nu$ (cm$^{-1}$) 3676 (NH), 3018, 2902, 1753 (C=O), 1508 (C=N), 1442, 1352, 754; $^1$H-NMR (DMSO-$d_6$): $\delta$ (ppm) 3.29 (t, $J$ = 7.0 Hz, 2H, –CH$_2$–CH$_2$–N), 4.11 (t, $J$ = 7.0 Hz, 2H, –CH$_2$–CH$_2$–N), 6.08 (s, 2H, -O-CH$_2$-O-), 6.86 (s, 1H, –N–CH=CH–N=), 6.94 (d, $J$ = 8.0 Hz, 1H, Ar–H), 7.10 (t, $J$ = 7.5 Hz, 1H, Ar–H), 7.25 (d, $J$ = 8.5 Hz, 1H, Ar–H), 7.27 (s, 1H, –N–CH=CH–N= ), 7.36 (t, $J$ = 7.5 Hz, 1H, Ar–H), 7.41 (s, 1H, Ar–H), 7.51 (d, $J$ = 7.5 Hz, 1H, Ar–H), 7.63 (s, 1H, –N–CH=N–), 8.25 (d,$J$ = 7.5 Hz, 1H, Ar–H), 9.12 (s, 1H, NH), 10.57 (s, 1H, NH); $^{13}$C-NMR (DMSO-$d_6$): $\delta$(ppm) 28.9 (C-2'), 42.8 (C-3'), 101.9 (–O–CH$_2$–O–),106.3, 108.5, (Ar–CH), 119.9 (C-5''), 121.1, 121.2, 122.9, 124.3, 128.4, 128.7, 129.6, 131.3, 135.7 (Ar-CH, Ar–C, C-4''), 137.8 (C-2''), 145.9, 148.3, 148.7, 153.4 (Ar–C, C=O, C=N); MS $m/z$ (ESI): 412.1 [M + H]$^+$, 413.0 [(M + 1) + H]$^+$, and 414.1 [(M + 2) + H]$^+$.

## 2.3. Crystal Structure Determination

Single crystals of the title compound 4 were achieved via slow evaporation of the ethyl alcohol solution of the pure compound at ambient temperature. Bruker APEX-II D8 Venture area diffractometer (Bruker, Billerica, MA, USA) implemented with graphite monochromatic Mo K$\alpha$radiation, $\lambda$ = 0.71073 Å at 293 (2) K, was employed for data collection. Bruker SAINT was harnessed for data reduction and cell refinement and SHELXT [24] was utilized for structure solving. Full-matrix least-square technique was utilized for final refinement with the aid of anisotropic thermal data for non-hydrogen atoms on $F$. CCDC 1,875,003 contains the supplementary crystallographic data for compound 4 can be obtained free of charge from the Cambridge Crystallographic Data Centre via www.ccdc.cam.ac.uk/data_request/cif.

## 2.4. Quantum Chemical Calculations

Quantum mechanical methods helped by the functional B3LYP/6-311+G were used to obtain the vibrational wavenumbers of the semicarbazone 4 and to optimizeits structure [25]. The theoretical investigations of the semicarbazone 4 were carried out with the aid of Gaussian 09W software [26]. The exchange and correlation functions embedded in the software with the occurrence of the density functional theory (DFT) are vital to get results closer to the experimental data. The distribution of assignments of the simulated wavenumbers were evaluated with the aid of vibrational energy distribution analysis (VEDA) 4 program [27].Crystal Explorer 3.1 was utilized to generate fingerprint plots and Hirshfeld surface map of the semicarbazone 4 [28]. NBO 3.1 program supported natural hybrid calculations, natural populations, and natural bonding orbital investigations [29].

## 2.5. Antifungal Activity

The MICs of the semicarbazone 4 towards various fungal strains were inspected according to literature method [20].

## 3. Results and Discussion

### 3.1. Chemistry

The synthetic procedure to prepare the target compound 4 is decorated in Scheme 1. The ketone 3 [19] was successfully obtained according to the previously reported method using acetophenone derivative **1** as a starting material. Subsequently, the semicarbazide N-(2-chlorophenyl)hydrazinecarboxamide [23] was allowed to react with compound **3** to yield the target semicarbazones **4** in an acceptable yield.

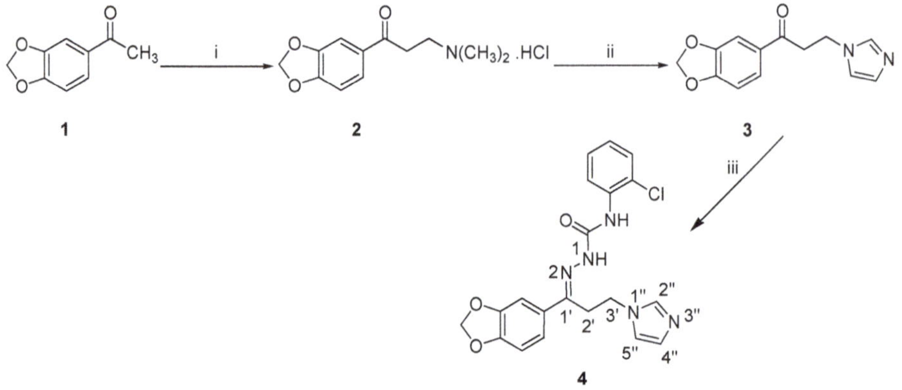

**Scheme 1.** Synthesis of the target semicarbazone **4**. *Reagents and conditions*: (i) $HN(CH_3)_2 \cdot HCl$, $(CH_2O)_n$, conc. HCl, ethanol, reflux, 2 h; (ii) Imidazole, water, reflux, 5 h; (iii) N-(2-Chlorophenyl)hydrazinecarboxamide, ethanol, acetic acid, rt, 18 h.

### 3.2. Crystal Structure of the Semicarbazone 4

In compound **4**, $C_{20}H_{18}ClN_5O_3$, the crystallographic information and refinement data are presented in Table 1. Table 2 illustrates the chosen bond angles and bond lengths. Figure 1 displayed the asymmetric unit which contains one independent molecule. All the bond angles and bond lengths occur in normal ranges [30]. The molecules are packed together in the crystal structure by one classical hydrogen bond and three non-classical hydrogen bonds, as presented in Figure 2 and Table 3. The 1,3-benzodioxole plane makes dihedral angles of 16.91° and 52.91° with a chlorophenyl ring and imidazole ring, respectively.

**Table 1.** X-ray experimental details of the semicarbazone **4**.

| | |
|---|---|
| **Crystal Data** | |
| Molecular formula | $C_{20}H_{18}ClN_5O_3$ |
| Mr | 411.84 |
| Crystal system, space group | Monoclinic, $P2_1/c$ |
| Temperature (K) | 293 |
| a, b, c (Å) | 8.7780 (6), 20.5417 (15), 11.0793 (9) |
| β (°) | 100.774 (2) |
| V (Å³) | 1962.5 (3) |
| Z | 4 |
| Radiation type | Mo Kα |
| μ (mm$^{-1}$) | 0.23 |
| Crystal size (mm³) | 0.33 × 0.19 × 0.11 |
| **Data Collection** | |
| Diffractometer | Brucker APEX-II D8 venture diffractometer |
| Absorption correction | Multi-scanSADABS Bruker 2014 |
| $T_{min}, T_{max}$ | 0.954, 0.962 |
| No. of measured, independent and observed [I> 2σ(I)] reflections | 28076, 4511, 2424 |
| $R_{int}$ | 0.095 |
| **Refinement** | |
| $R[F^2 > 2\sigma(F^2)], wR(F^2), S$ | 0.060, 0.137, 1.04 |
| No. of reflections | 4511 |
| No. of parameters | 270 |
| No. of restraints | 0 |
| H-atom treatment | H atoms treated by a mixture of independent and constrained refinement |
| $\Delta\rho_{max}, \Delta\rho_{min}$ (e Å$^{-3}$) | 0.18, −0.29 |

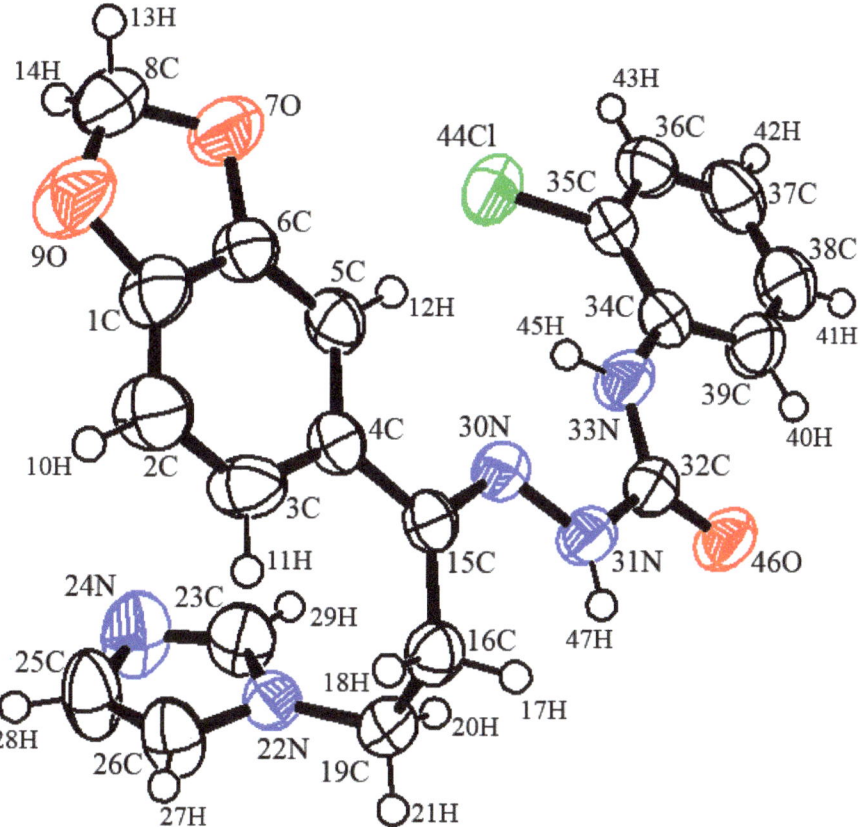

**Figure 1.** Oak ridge thermal ellipsoid plot diagram of the semicarbazones **4** drawn at 50% ellipsoids for non-hydrogen atoms.

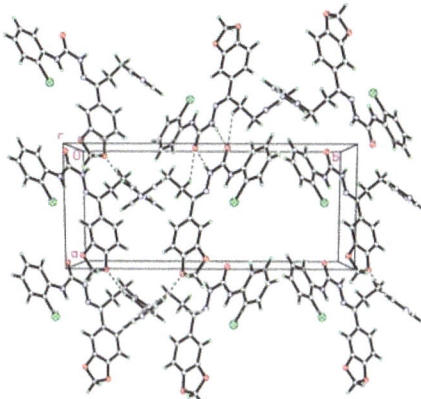

**Figure 2.** Molecular packing of the semicarbazone **4** manifesting hydrogen bonds which are drawn as dashed lines.

**Table 2.** Geometric parameters (Å, °) of the semicarbazone **4**.

| Atom | Bond Length/Angle | Atom | Bond Length/Angle |
|---|---|---|---|
| Cl44–C35 | 1.736 (3) | N22–C26 | 1.359 (3) |
| O7–C6 | 1.377 (3) | N24–C23 | 1.300 (4) |
| O7–C8 | 1.416 (3) | N24–C25 | 1.350 (4) |
| O9–C1 | 1.372 (3) | N30–N31 | 1.378 (3) |
| O9–C8 | 1.422 (4) | N30–C15 | 1.288 (3) |
| O46–C32 | 1.224 (3) | N31–C32 | 1.360 (3) |
| N22–C19 | 1.454 (3) | N33–C32 | 1.361 (3) |
| N22–C23 | 1.351 (3) | N33–C34 | 1.403 (3) |
| C6–O7–C8 | 105.8 (2) | N30–C15–C4 | 116.0 (2) |
| C1–O9–C8 | 105.4 (2) | N30–C15–C16 | 123.5 (2) |
| C19–N22–C23 | 126.9 (2) | N22–C19–C16 | 114.3 (2) |
| C19–N22–C26 | 127.8 (2) | N22–C23–N24 | 112.9 (3) |
| C23–N22–C26 | 105.3 (2) | N24–C25–C26 | 111.2 (3) |
| C23–N24–C25 | 104.3 (3) | N22–C26–C25 | 106.4 (2) |
| N31–N30–C15 | 119.1 (2) | N31–C32–N33 | 114.8 (2) |
| N30–N31–C32 | 119.4 (2) | O46–C32–N31 | 120.6 (2) |
| C32–N33–C34 | 127.9 (2) | O46–C32–N33 | 124.6 (3) |
| O7–C6–C1 | 109.2 (2) | N33–C34–C39 | 124.1 (3) |
| O7–C6–C5 | 128.4 (2) | N33–C34–C35 | 118.4 (2) |
| O9–C1–C2 | 128.8 (3) | Cl44–C35–C34 | 119.5 (2) |
| O9–C1–C6 | 109.9 (2) | Cl44–C35–C36 | 118.8 (2) |
| O7–C8–O9 | 107.5 (2) | | |

**Table 3.** Hydrogen-bond geometry (Å) of the semicarbazone **4**.

| D–H···A | D–H | H···A | D···A | D–H···A |
|---|---|---|---|---|
| N31–H4N31···O46i | 0.85(3) | 2.06(3) | 2.896(3) | 169(3) |
| C16–H17B···O46i | 0.9700 | 2.5100 | 3.207(3) | 129.00 |
| C23–H29A···O9ii | 0.9300 | 2.4500 | 3.301(4) | 152.00 |
| C26–H27A···N24iii | 0.9300 | 2.4900 | 3.326(4) | 150.00 |
| C39–H40A···O46 | 0.9300 | 2.3000 | 2.886(4) | 121.00 |

Symmetry codes: (i) −x−2, −y−1, −z−1; (ii) x−1, y, z; (iii) x, −y−3/2, z−1/2.

### 3.3. Structural Geometry Analysis

The geometry of the semicarbazone **4** was optimized with the aid of DFT theory at B3LYP/6-311++G (d,p) level of basis set by Gaussian 09 program package. The atom numbering scheme is displayed in Figure 3. No constraints on the bond lengths, bond angles or dihedral angles were observed in the calculations and all atoms were free to optimize. Table 4 illustrates the relevant optimized geometrical parameters for the semicarbazone **4** along with its experimental values. There is a difference in values from crystallographic data which might be arise from the fact that theoretical results belong to gaseous phase (isolated molecule), whereas the experimental results belong to solid phase. A statistical correlation graph showing the gaseous and solid phase bond lengths is depicted in Figure S1 with $R^2$ value of 0.97233.

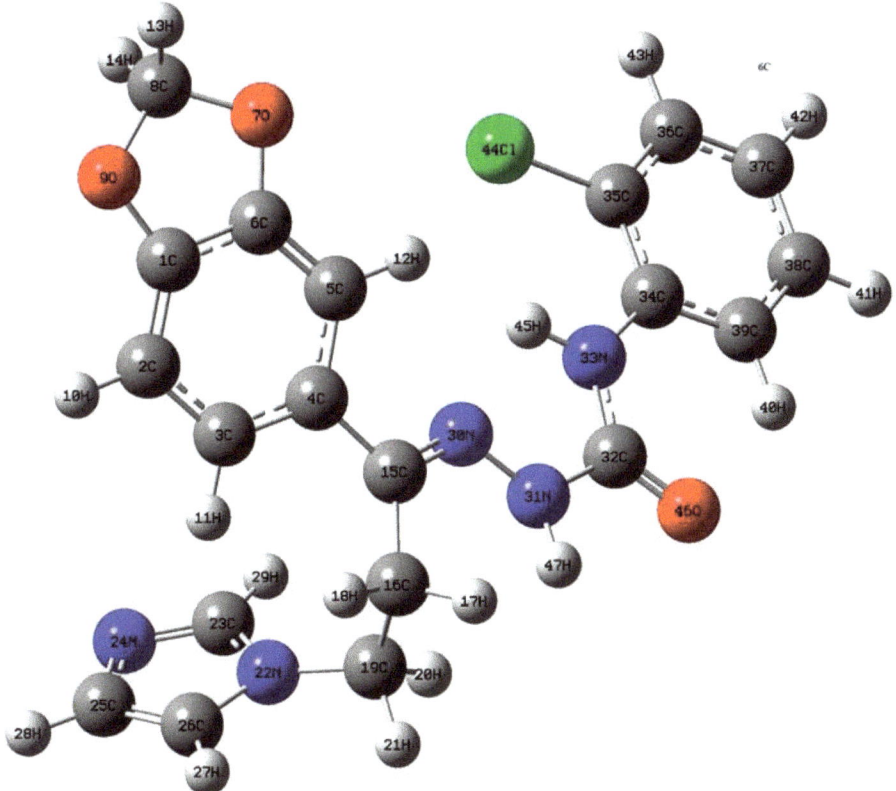

**Figure 3.** Atom numbering scheme of optimized semicarbazone 4 structure.

The title molecule bears a chlorophenyl moiety connected with a benzodioxole ring through a hydrazinecarboxamide bridge. Connection was extended to theimidazole ring through an ethylene group. Computational geometry optimization of the title molecule revealed that the global energy minimum for the non-planar structure was obtained by 1732.9189 Hartree. The calculated bond lengths of C3–H11, C2–H10, C5–H12 of the benzodioxole ring are 1.0816, 1.0822, and 1.0811 Å, respectively. The shortening of C5–H12 might beowing to the presence of intramolecular C5–H12 $\cdots$ N30 hydrogen bonding and it is significantly shorter than the van der Waals radii (2.4104 Å).

The angle of 95.83° between C5–H12 $\cdots$ N30 is also within the angle limit manifesting C–H $\cdots$ N intramolecular hydrogen bonding. The C–N bond lengths C32–N31 (1.4042 Å) and C32–N33 (1.3875 Å) are shorter than the normal C–N bond length of 1.480 Å. This discrepancy could be attributed to the conjugation of Pi-type electrons of the carbonyl group and nitrogen atom, allowing the electrons to smear out along the C–N bond. Table 4 displayed bond lengths of C3–C4 and C4–C5 in the benzodioxole ring as well as C34–C35, C34–C39 in the chlorophenyl ring which are significantly longer than the other C–C bond lengths in both rings. This is due to the intramolecular charge transfer (ICT) between two rings through hydrazinecarboxamide moiety. DFT calculations indicates a decrease in the endocycyclic angles C3–C4–C5 and C35–C34–C39 by 0.76° and 2.42°, respectively, from the normal endocyclic angle of 120° supporting intramolecular charge transfer interactions.

Table 4. Selected optimized parameters for the semicarbazone 4.

| Parameters | DFT | Exp. | Parameters | DFT | Exp. |
|---|---|---|---|---|---|
| Bond lengths\Å | | | Dihedral Angles\° | | |
| C1–C2 | 1.3751 | 1.356 | C5–C4–C15–N30 | −0.51 | −1.54 |
| C1–C6 | 1.3954 | 1.371 | C3–C4–C15–N30 | 179.97 | 178.01 |
| C2–C3 | 1.4053 | 1.392 | C15–N30–N31–C32 | 174.78 | 176.1 |
| C2–H10 | 1.0822 | 0.931 | N30–N31–C32–N33 | −174.29 | −2.89 |
| C3–C4 | 1.4011 | 1.378 | N32–C32–N33–C34 | −178.49 | 176.6 |
| C3–H11 | 1.0816 | 0.931 | C32–N33–C34–C39 | −1.04 | −1.54 |
| C4–C5 | 1.4189 | 1.378 | C32–N33–C34–C35 | 179.15 | 165.48 |
| C4–C15 | 1.4852 | 1.480 | C4–C15–C16–C19 | −88.51 | 101.58 |
| C5–C6 | 1.3701 | 1.359 | C15–C16–C19–N22 | −175.16 | −76.52 |
| C5–H12 | 1.0811 | 0.930 | C16–C19–N22–C26 | −73.22 | −78.94 |
| C15=N30 | 1.2889 | 1.2881 | C16–C19–N22–C23 | 103.09 | 101.11 |
| N30–N31 | 1.3524 | 1.3781 | Bond Angles\° | | |
| C32–N31 | 1.4042 | 1.378 | C5–C4–C15 | 121.79 | 121.74 |
| C32–N33 | 1.3875 | 1.402 | C3–C4–C15 | 118.89 | 119.62 |
| C34–N33 | 1.4016 | 1.362 | C3–C4–C5 | 119.33 | 122.40 |
| C34–C35 | 1.4072 | 1.396 | N33–C34–C35 | 118.58 | 118.36 |
| C34–C39 | 1.4035 | 1.389 | N33–C34–C39 | 123.84 | 124.08 |
| C35–C36 | 1.3873 | 1.378 | C35–C34–C39 | 117.58 | 117.56 |
| C36–C37 | 1.3918 | 1.368 | | | |
| C37–C38 | 1.3925 | 1.370 | | | |
| C38–C39 | 1.3908 | 1.380 | | | |
| C39–C34 | 1.4035 | 1.389 | | | |

### 3.4. Natural Bond Orbital Analysis

The NBO calculations were performed using the NBO 3.1 program [29] as implemented in the Gaussian 09 package at the DFT level. It is proved that NBO analysis is an efficient approach for the chemical interpretation of hyperconjugative interactions within the molecule. Additionally, it can be used to explain the electron density transfer (EDT) from the filled lone electron pairs of n(Y) of the "Lewis base" Y into the unfilled antibond σ*(X–H) of the "Lewis acid" X–H in X–H⋯Y hydrogen bonding systems. In addition, NBO analysis can be exploited to elucidate inter and intramolecular hydrogen bonding as well as intermolecular charge transfer in organic molecules.

The intramolecular C–H⋯N hydrogen bonding was formed due to the overlap between n1(O) and σ*C–H which results in intramolecular charge transfer (ICT) causing stabilization of the system. These interactions lead to an increased electron density of the C–H antibonding orbital, which strengthens the C–H bond. In addition, NBO analyses confirmed C5–H12⋯N30 intramolecular hydrogen bonding formed by the overlap between a lone pair n1(N30) and σ*C5–H12 antibonding orbitals with stabilization energy of 4.40 kcal/mol. Other important interactions are displayed in Table 5.

### 3.5. Natural Population Analysis

Atomic charges and electron distribution within a molecule can be effectively calculated with the aid of natural population analysis (NPA) [31]. Figure 4 clearly illustrates the atomic charges of the semicarbazone 4 obtained by NPA. The hydrogen atoms of the semicarbazone 4 exhibited net positive charges. The atoms H45 and H47 manifested more positive charge than the other hydrogen atoms owing to their connection to nitrogen atoms. Among the hydrogen atoms of the chlorophenyl and benzodioxole ring, H12 showed the highest positive charge, being involved in C–H⋯N intramolecular hydrogen bonding. Moreover, the all carbon atoms manifested negative charges except C1, C6, C8, C15, C23, C32, and C34 owing to their connection to nitrogen or oxygen atoms. These findings suggest that the N and O atoms are the preferred sites for protonation. C32 is the most positively charged atom, while N33 is the most negatively charged one. On the other hand, C35 possesses the lowest negative charge, as compared with the other carbon atom in the phenyl moiety owing to its connection to the chlorine atom.

Table 5. Second-order perturbation theory analysis of Fock matrix in NBO basis for the semicarbazone 4.

| Donor (i) | | Acceptor (j) | | E(2) [a] kcal/mol | E(j)-E(i) [b] (a.u.) | F(i,j) [c] (a.u.) |
|---|---|---|---|---|---|---|
| NBO | Occupancy | NBO | Occupancy | | | |
| σ(C1–C2) | 1.97467 | σ*(C1–C6) | 0.03968 | 4.48 | 1.27 | 0.068 |
| σ(C1–C2) | 1.97467 | σ*(C1–O9) | 0.02853 | 3.66 | 0.34 | 0.032 |
| σ(C1–C2) | 1.97467 | σ*(C1–O7) | 0.03253 | 1.47 | 1.07 | 0.035 |
| σ (C1–C6) | 1.97447 | σ*(C1–C2) | 0.02065 | 4.65 | 1.35 | 0.071 |
| σ (C1–C6) | 1.97447 | σ*(C15–N30) | 0.01622 | 1.51 | 0.76 | 0.032 |
| σ (C1–O9) | 1.98708 | σ*(C35–Cl44) | 0.03389 | 0.53 | 3.49 | 0.039 |
| σ(C2–H10) | 1.97906 | σ*(C37–H42) | 0.01319 | 0.58 | 0.79 | 0.019 |
| σ(C3–C4) | 1.97383 | σ*(C1–O9) | 0.02853 | 12.75 | 0.31 | 0.056 |
| σ(C3–C4) | 1.97383 | σ*(C2–C3) | 0.02065 | 6.45 | 0.17 | 0.078 |
| σ(C3–C4) | 1.97383 | σ*(N24–C25) | 0.01083 | 38.23 | 0.18 | 0.074 |
| π(C3–C4) | 1.69506 | π*(C1–C2) | 0.02065 | 0.56 | 3.68 | 0.041 |
| π(C3–C4) | 1.69506 | π*(C25–H28) | 0.01691 | 33.36 | 4.07 | 0.356 |
| π(C5–C6) | 1.70385 | π*(C1–C2) | 0.02065 | 14.57 | 0.47 | 0.075 |
| π(C5–C6) | 1.70385 | π*(C34–C35) | 0.46482 | 0.66 | 0.45 | 0.016 |
| n1(N30) | 1.90915 | σ*(C5–H12) | 0.01401 | 4.40 | 0.82 | 0.055 |
| n1(N33) | 1.70147 | σ*(N31–C32) | 0.08922 | 13.15 | 0.46 | 0.069 |

[a]: E(2) means energy of stabilization interactions; [b]: Energy difference between donor-to-acceptor, i and j NBO orbitals; [c]: F(i,j) is the Fock matrix element between i and j NBO orbitals.

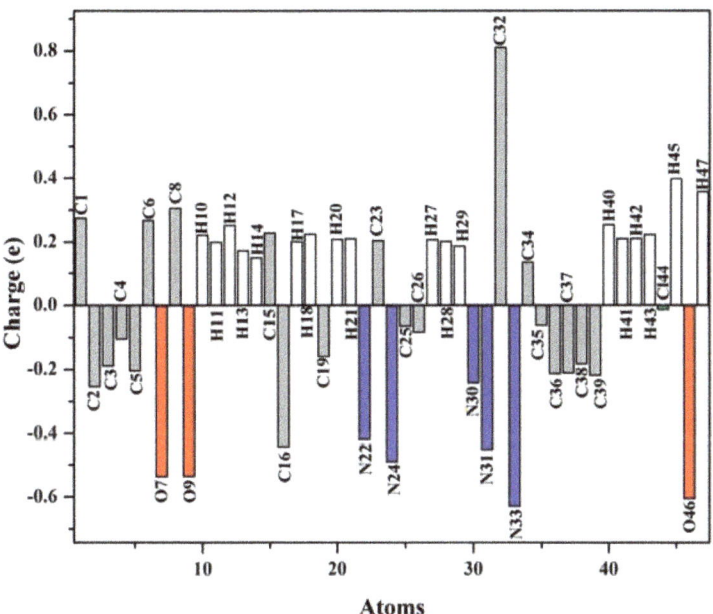

Figure 4. Charge distribution on the atoms of the semicarbazone 4.

### 3.6. Frontier Molecular Orbital Analysis

The kinetic stability and chemical reactivity of a molecule could be examined through the energy gap between its HOMO (highest occupied molecular orbital) and its LUMO (lowest unoccupied molecular orbital) [32]. The B3LYP/6-311++G(d,p) level of theory was utilized to calculate the energy gap for the semicarbazone 4 and Figure 5 illustrates the graphical representation of its frontier molecular orbitals. The electron distribution is scattered in HOMO over the benzodioxole ring, the hydrazinecarboxamide moiety, and is slightly upon the chlorophenyl ring, whereas the LUMO is mainly disseminated over the benzodioxole fragment. The electron density is transferred in the HOMO→LUMO transition from the chlorophenyl fragment to the benzodioxole moiety through the hydrazine carboxamide bridge. This transition illustrates the charge transfer from the electron-donor functionalities to electron-acceptor fragments via a π-conjugated path. The FMO (frontier molecular orbital) energy gap of the semicarbazone 4 was found to be 4.23 eV (HOMO energy = −6.007 eV, LUMO energy = −1.777 eV). This energy gap leads to charge transfer within the semicarbazone 4, which could promote its bioactivity.

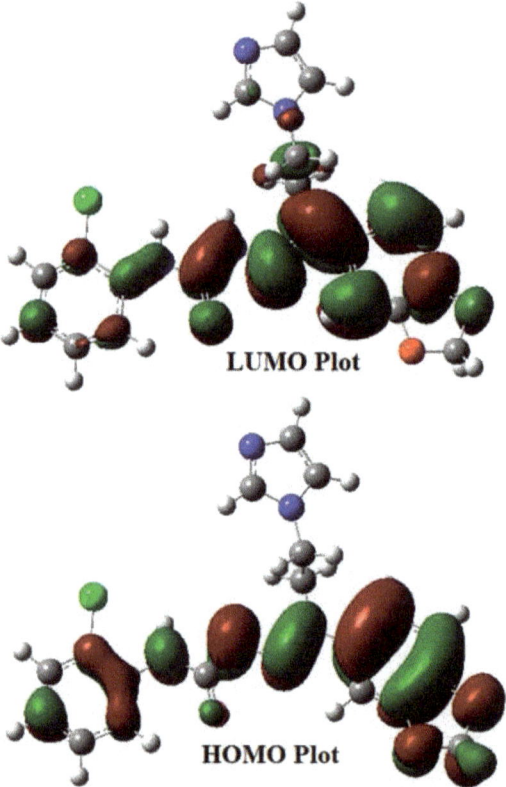

**Figure 5.** Frontier molecular orbitals of the semicarbazone 4.

*3.7. Molecular Docking Simulations*

Gaussian 09 program was used to optimize the structure of the semicarbazones 4 with the aid of the density functional theory using [26]. AutoDock Tools-1.5.4 implemented in MGL Tools-1.5.4 package was harnessed to carry out molecular docking investigations [33]. The fungal RNA Kinase (PDB ID: 5U32) target protein was chosento perform the current docking analysis [34]. The three-dimensional (3D) coordinates file of 5U32 with a resolution of 2.27 Å was secured from RCSB (Research Collaboratory for Structural Bioinformatics) protein data bank [35]. Affinity grids centered on the active site with 126×126×126 grid size with a spacing of 0.42 Å were generated with the aid of AutoGrid 4.2 [36]. The docking procedure was carried out as described in the literature [21]. The docking results were examined through sorting the binding free energy envisaged by the docked confirmations of the semicarbazone 4. The presaged best confirmation binding energy was −4.02 kcal/mol. The amino acids PHE436, LEU608, ASP607, and LYS610 in the active site of the target protein binds with the ligand by hydrogen bonding with bond lengths of 1.83 and 2.26 as well as 2.04, 2.27, and 1.92, respectively. The protein-ligand interaction complex is given in Figure 6, suggesting the possible binding mode of the title compound 4 to the target protein 5U32.

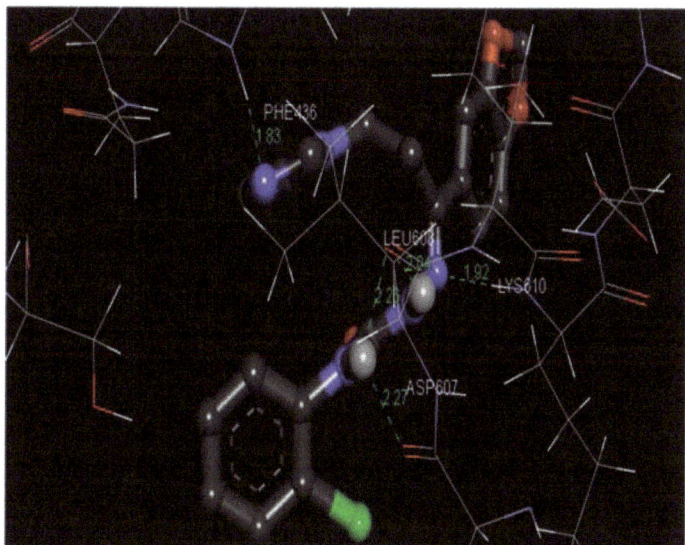

**Figure 6.** Pose of the semicarbazone 4 with the amino acid residues of the target protein.

*3.8. Hirshfeld Surface Analysis*

The Hirshfeld surfaces and their associated 2D fingerprint plots for the title compound 4 were computed, taking single crystal X-ray crystal structure as input. The Hirshfeld surface emerged from an attempt to define the space occupied by a molecule in a crystal for the purpose of partitioning the crystal electron density into molecular fragments [37]. It provides a 3D picture of close contacts in a crystal, and these contacts can be summarized in a fingerprint plot [38]. For each point on the Hirshfeld surface, two distances are defined, namely the $d_e$: the distance from a point to the nearest nucleus external to the surface and the $d_i$: the distance to the nearest nucleus internal to the surface. A plot of $d_i$ versus $d_e$ is a 2D fingerprint plot which recognizes the existence of different types of intermolecular interactions. The Hirshfeld surfaces of the title compound are given in Figure 7 and it explains the surfaces that have been mapped over $d_{norm}$, $d_e$, and $d_i$ and curvedness (3D plots).

The function $d_{norm}$ (normalized chemical contacts) is a ratio encompassing the distances of any surface point to the nearest $d_i$ and $d_e$ atom and the van der Waals radii of the atoms. The value of $d_{norm}$ may be positive or negative depending on whether the intermolecular contacts being either longer or shorter than the van der Waals separations. The negative value of $d_{norm}$ indicates the sum of $d_i$ and $d_e$ is shorter than the sum of the relevant van der Waals radii, which is considered to be the closest contact and is visualized as red color in the Hirshfeld surfaces. The blue color denotes contacts longer than the sum of van der Waals radii with positive $d_{norm}$ values, whereas contacts close to van der Waalsradiiwith $d_{norm}$ equal to zero are colored white. The intermolecular interactions outwards the H⋯H/N⋯H/C⋯H/O⋯H bonds as well as the overall fingerprint region of the title molecule are displayed in Figure 8. With these analyses, the division of contributions is possible for different interactions including H⋯H, N⋯H, C⋯H, and O⋯H, which commonly overlap in the full finger print plots. Figure 7 shows the $d_{norm}$ surface of the semicarbazone 4, highlighting only H⋯H/N⋯H intermolecular contacts. These interactions comprise 35.5% and 5.7% of the total Hirshfeld surface area for this molecule, respectively. O⋯H/N⋯H intermolecular interactions are represented by a two sharp spike in the 2D fingerprint plot (Figure 8). The relative contributions of the C$\cdots$H and O$\cdots$H contacts are 12.2% and 7.5%, respectively.

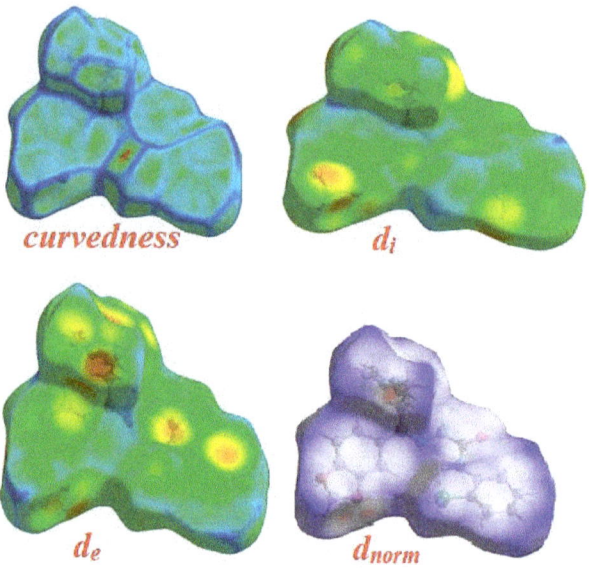

**Figure 7.** Hirshfeld surfaces for $d_{norm}$, $d_i$, and $d_e$ and curvedness for thesemicarbazone **4**.

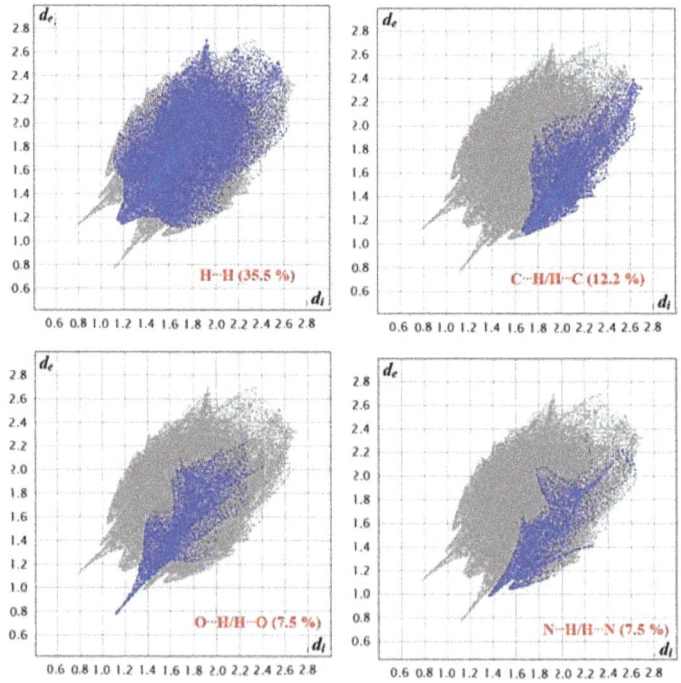

**Figure 8.** Two dimension finger print plots showing various intermolecular contacts in thesemicarbazones **4**.

## 3.9. Antifungal Activity of the Semicarbazone 4

Table 6 illustrates the minimum inhibitory concentrations (MICs) values of the antifungal profile of the target semicarbazone **4**, as well as the reference standard antifungal drugs, fluconazole, and ketoconazole. The best antifungal activity of the semicarbazone **4** was displayed against *C. albicans* with MIC value of 0.156 µmol/mL.

**Table 6.** The minimum inhibitory concentrations (MICs) values for the antifungal activity of the semicarbazone **4**, fluconazole and ketoconazole towards *Aspergillus niger* and various *Candida* species.

| Compound No. | MIC (µmol/mL) | | | |
|---|---|---|---|---|
| | *C. albicans* | *C. tropicalis* | *C. parapsilosis* | *Aspergillus niger* |
| 4 | 0.156 | > 1.24 | 0.311 | 0.622 |
| Fluconazole | 0.051 | 0.045 | 0.047 | ND |
| Ketoconazole | ND | ND | ND | 0.02 |

ND: not determined.

## 4. Conclusions

(2*E*)-2-[1-(1,3-Benzodioxol-5-yl)-3-(1*H*-imidazol-1-yl)propylidene]-*N*-(2-chlorophenyl)hydrazinecarboxamide (**4**) has been synthesized and identified with different spectroscopic techniques. The assigned chemical structure of the title compound **4** as well as the (*E*)-configuration of its imine functionality were unambiguously assured using single crystal X-ray analysis.

The optimized geometry of the semicarbazone **4** revealed that the predicted values are consistent with the experimental results with few exceptions that are insignificant. The obtained theoretical model of the title compound **4** by DFT calculations describes the possibility of C–H $\cdots$ N intramolecular hydrogen bonding and ICT charge transfer interactions that stabilize its structure. These results were further confirmed via NBO and NPA analyses on the title compound **4**. Frontier molecular orbital analysis also confirmed the presence of intramolecular charge transfer within the title molecule **4**, leading to its stability and henceits bioactivity. Hirshfeld surface analysis revealed the occurrence of strong and weak intermolecular interactions in the crystalline state of the semicarbazone **4** like O$\cdots$H, C$\cdots$H and N$\cdots$H bonding. The in vitro antifungal potential of the title compound **4** was examined against different fungal strains and the best activity was manifested against *C. albicans* with MIC value of 0.156 µmol/mL. A molecular docking study suggested the anticipated binding mode of the title compound **4** to its target protein. The results of the current study inspired us to design new antifungal candidates bearing the scaffold of the title compound **4** with the hope of getting more potent antifungal drug-like surrogates with better antifungal profile.

**Supplementary Materials:** The following are available online at http://www.mdpi.com/2073-4352/9/2/82/s1, Figure S1: Correlation plot for calculated and experimental bond lengths of semicarbazone 4; Figure S2: $^1$H NMR spectrum of the semicarbazone 4; Figure S3: Enlarged part of the $^1$H NMR spectrum of the semicarbazone 4; Figure S4: $^{13}$C NMR spectrum of the semicarbazone 4.

**Author Contributions:** R.I.A.-W. and A.R.A.-G. synthesized and characterized the title molecule. H.A.G. conducted X-ray analysis. M.H.A.-A. conducted the antifungal profiling. S.A.V.A. and I.H.J. performed the computational work. M.I.A. conceptualized the work and prepared the manuscript for publication. All authors discussed the contents of the manuscript.

**Funding:** The authors would like to extend their sincere appreciation to the Deanship of Scientific Research at King Saud University for its funding of this research through the Research Group Project no. RGP-196.

**Conflicts of Interest:** The authors have declared that there is no conflict of interests.

## References

1. Crunkhorn, S. Fungal infection: Protecting from *Candida albicans*. *Nat. Rev. Drug Discov.* **2016**, *15*, 604. [CrossRef] [PubMed]

2. Vandeputte, P.; Ferrari, S.; Coste, A.T. Antifungal resistance and new strategies to control fungal infections. *Int. J. Microbiol.* **2011**, *2012*, 713687. [CrossRef] [PubMed]
3. Wu, J.; Ni, T.; Chai, X.; Wang, T.; Wang, H.; Chen, J.; Jin, Y.; Zhang, D.; Yu, S.; Jiang, Y. Molecular docking, design, synthesis and antifungal activity study of novel triazole derivatives. *Eur. J. Med. Chem.* **2018**, *143*, 1840–1846. [CrossRef] [PubMed]
4. Lino, C.I.; de Souza, I.G.; Borelli, B.M.; Matos, T.T.S.; Teixeira, I.N.S.; Ramos, J.P.; de Souza Fagundes, E.M.; de Oliveira Fernandes, P.; Maltarollo, V.G.; Johann, S. Synthesis, molecular modeling studies and evaluation of antifungal activity of a novel series of thiazole derivatives. *Eur. J. Med. Chem.* **2018**, *151*, 248–260. [CrossRef] [PubMed]
5. Sanglard, D. Emerging threats in antifungal-resistant fungal pathogens. *Front. Med.* **2016**, *3*, 11. [CrossRef] [PubMed]
6. Pappas, P.G.; Kauffman, C.A.; Andes, D.R.; Clancy, C.J.; Marr, K.A.; Ostrosky-Zeichner, L.; Reboli, A.C.; Schuster, M.G.; Vazquez, J.A.; Walsh, T.J. Clinical practice guideline for the management of candidiasis: 2016 update by the Infectious Diseases Society of America. *Clin. Infect. Dis.* **2015**, *62*, e1–e50. [CrossRef]
7. Aoyama, Y.; Yoshida, Y.; Sato, R. Yeast cytochrome P-450 catalyzing lanosterol 14 alpha-demethylation. II. Lanosterol metabolism by purified P-450 (14) DM and by intact microsomes. *J. Biol. Chem.* **1984**, *259*, 1661–1666.
8. Aboul-Enein, M.N.; El-Azzouny, A.A.; Attia, M.I.; Saleh, O.A.; Kansoh, A.L. Synthesis and anti-*Candida* potential of certain novel 1-[(3-substituted-3-phenyl)propyl]-1*H*-imidazoles. *Arch. Pharm.* **2011**, *344*, 794–801. [CrossRef]
9. Roman, G.; Mares, M.; Nastasa, V. A novel antifungal agent with broad spectrum: 1-(4-biphenylyl)-3-(1*H*-imidazol-1-yl)-1-propanone. *Arch. Pharm.* **2013**, *346*, 110–118. [CrossRef]
10. Attia, M.I.; Radwan, A.A.; Zakaria, A.S.; Almutairi, M.S.; Ghoneim, S.W. 1-Aryl-3-(1*H*-imidazol-1-yl)propan-1-ol esters: Synthesis, anti-*Candida* potential and molecular modeling studies. *Chem. Cent. J.* **2013**, *7*, 168. [CrossRef]
11. Venkatachalam, T.K.; Bernhardt, P.V.; Noble, C.J.; Fletcher, N.; Pierens, G.K.; Thurecht, K.J.; Reutens, D.C. Synthesis, characterization and biological activities of semicarbazones and their copper complexes. *J. Inorg. Biochem.* **2016**, *162*, 295–308. [CrossRef] [PubMed]
12. Beraldo, H.; Gambino, D. The wide pharmacological versatility of semicarbazones, thiosemicarbazones and their metal complexes. *Mini Rev. Med. Chem.* **2004**, *4*, 31–39. [PubMed]
13. Jafri, L.; Ansari, F.L.; Jamil, M.; Kalsoom, S.; Qureishi, S.; Mirza, B. Microwave-assisted synthesis and bioevaluation of some semicarbazones. *Chem. Biol. Drug Des.* **2012**, *79*, 950–959. [CrossRef] [PubMed]
14. Laly, S.; Parameswaran, G. Synthesis and characterisation of some thiosemicarbazone complexes of Ag (I), Pt (II) and Pd (II). *Asian J. Chem.* **1993**, *5*, 712–718.
15. Aboul-Enein, M.N.; El-Azzouny, A.A.; Attia, M.I.; Maklad, Y.A.; Amin, K.M.; Abdel-Rehim, M.; El-Behairy, M.F. Design and synthesis of novel stiripentol analogues as potential anticonvulsants. *Eur. J. Med. Chem.* **2012**, *47*, 360–369. [CrossRef] [PubMed]
16. Leite, A.C.L.; da Silva, K.P.; de Souza, I.A.; de Araújo, J.M.; Brondani, D.J. Synthesis, antitumour and antimicrobial activities of new peptidyl derivatives containing the 1,3-benzodioxole system. *Eur. J. Med. Chem.* **2004**, *39*, 1059–1065. [CrossRef] [PubMed]
17. Attia, M.I.; El-Brollosy, N.R.; Kansoh, A.L.; Ghabbour, H.A.; Al-Wabli, R.I.; Fun, H.-K. Synthesis, single crystal X-ray structure, and antimicrobial activity of 6-(1,3-benzodioxol -5-ylmethyl)-5-ethyl-2-{[2-(morpholin-4-yl)ethyl]sulfanyl}pyrimidin-4(3*H*)-one. *J. Chem.* **2014**, *2014*, 457430. [CrossRef]
18. Attia, M.I.; Zakaria, A.S.; Almutairi, M.S.; Ghoneim, S.W. In vitro anti-*Candida* activity of certain new 3-(1*H*-imidazol-1-yl)propan-1-one oxime esters. *Molecules* **2013**, *18*, 12208–12221. [CrossRef] [PubMed]
19. Al-Wabli, R.I.; Al-Ghamdi, A.R.; Ghabbour, H.A.; Al-Agamy, M.H.; Monicka, J.C.; Joe, I.H.; Attia, M.I. Synthesis, X-ray single crystal structure, molecular docking and DFT computations on *N*-[(1*E*)-1-(2*H*-1,3-benzodioxol-5-yl)-3-(1*H*-imidazol-1-yl)propylidene]-hydroxylamine: A new potential antifungal agent precursor. *Molecules* **2017**, *22*, 373. [CrossRef]
20. Al-Wabli, R.I.; Al-Ghamdi, A.R.; Ghabbour, H.A.; Al-Agamy, M.H.; Attia, M.I. Synthesis, single crystal X-ray analysis, and antifungal profiling of certain new oximino ethers bearing imidazole nuclei. *Molecules* **2017**, *22*, 1895. [CrossRef]

21. Al-Wabli, R.I.; Al-Ghamdi, A.R.; Primsa, I.; Ghabbour, H.A.; Al-Agamy, M.H.; Joe, I.H.; Attia, M.I. (2E)-2-[1-(1,3-benzodioxol-5-yl)-3-(1H-imidazol-1-yl)propylidene]-N-(4-methoxyphenyl)hydrazine carboxamide: Synthesis, crystal structure, vibrational analysis, DFT computations, molecular docking and antifungal activity. *J. Mol. Struct.* **2018**, *1166*, 121–130. [CrossRef]
22. Al-Wabli, R.I.; Al-Ghamdi, A.R.; Ghabbour, H.A.; Al-Agamy, M.H.; Attia, M.I. Synthesis and spectroscopic identification of certain imidazole-semicarbazone conjugates bearing benzodioxole moieties: New antifungal agents. *Molecules* **2019**, *24*, 200. [CrossRef] [PubMed]
23. Beukers, M.W.; Wanner, M.J.; Von Frijtag Drabbe Künzel, J.K.; Klaasse, E.C.; Ijzerman, A.P.; Koomen, G.-J. N-6-Cyclopentyl-2-(3-phenylaminocarbonyltriazene-1-yl)adenosine (TCPA), a very selective agonist with high affinity for the human adenosine A1 receptor. *J. Med. Chem.* **2003**, *46*, 1492–1503. [CrossRef] [PubMed]
24. Sheldrick, G.M. A short history of SHELX. *Acta Cryst. A* **2008**, *64*, 112–122. [CrossRef] [PubMed]
25. Becke, A.D. Density-functional thermochemistry. III. The role of exact exchange. *J. Chem. Phys.* **1993**, *98*, 5648–5652. [CrossRef]
26. Frisch, M.J.; Trucks, G.W.; Schlegel, H.B.; Scuseria, G.E.; Robb, M.A.; Cheeseman, J.R.; Scalmani, G.; Barone, V.; Mennucci, B.; Petersson, G.A.; et al. *Gaussian-09*; Revision A.02; Gaussian, Inc.: Wallingford, CT, USA, 2009.
27. Jamróz, M.H. Vibrational energy distribution analysis (VEDA): Scopes and limitations. *Spectrochim. Acta Part A Mol. Biomol. Spectrosc.* **2013**, *114*, 220–230. [CrossRef] [PubMed]
28. Wolff, S.; Grimwood, D.; McKinnon, J.; Jayatilaka, D.; Spackman, M. *Crystal Explorer 2.0*; University of Western Australia: Perth, Australia, 2007.
29. Reed, A.E.; Curtiss, L.A.; Weinhold, F. Intermolecular interactions from a natural bond orbital, donor-acceptor viewpoint. *Chem. Rev.* **1988**, *88*, 899–926. [CrossRef]
30. Allen, F.H.; Kennard, O.; Watson, D.G.; Brammer, L.; Orpen, A.G.; Taylor, R. Tables of bond lengths determined by X-ray and neutron diffraction. Part 1. Bond lengths in organic compounds. *J. Chem. Soc. Perkin Trans. 2* **1987**, S1–S19. [CrossRef]
31. Reed, A.E.; Weinstock, R.B.; Weinhold, F. Natural population analysis. *J. Chem. Phys.* **1985**, *83*, 735–746. [CrossRef]
32. Fleming, I. *Frontier Orbitals and Organic Chemical Reactions*; Wiley: Hoboken, NJ, USA, 1977.
33. Morris, G.M.; Huey, R.; Lindstrom, W.; Sanner, M.F.; Belew, R.K.; Goodsell, D.S.; Olson, A.J. AutoDock4 and AutoDockTools4: Automated docking with selective receptor flexibility. *J. Comp. Chem.* **2009**, *30*, 2785–2791. [CrossRef]
34. Bonanno, J.B.; Edo, C.; Eswar, N.; Pieper, U.; Romanowski, M.J.; Ilyin, V.; Gerchman, S.E.; Kycia, H.; Studier, F.W.; Sali, A. Structural genomics of enzymes involved in sterol/isoprenoid biosynthesis. *Proc. Natl. Acad. Sci. USA* **2001**, *98*, 12896–12901. [CrossRef] [PubMed]
35. Bernstein, F.C.; Koetzle, T.F.; Williams, G.J.; Meyer, E.F., Jr.; Brice, M.D.; Rodgers, J.R.; Kennard, O.; Shimanouchi, T.; Tasumi, M. The protein data bank: A computer-based archival file for macromolecular structures. *Arch. Biochem. Biophys.* **1978**, *185*, 584–591. [CrossRef]
36. Morris, G.M.; Goodsell, D.S.; Halliday, R.S.; Huey, R.; Hart, W.E.; Belew, R.K.; Olson, A.J. Automated docking using a Lamarckian genetic algorithm and an empirical binding free energy function. *J. Comp. Chem.* **1998**, *19*, 1639–1662. [CrossRef]
37. Spackman, M.A.; Byrom, P.G. A novel definition of a molecule in a crystal. *Chem. Phys. Lett.* **1997**, *267*, 215–220. [CrossRef]
38. Spackman, M.A.; Jayatilaka, D. Hirshfeld surface analysis. *Cryst. Eng. Comm.* **2009**, *11*, 19–32. [CrossRef]

© 2019 by the authors. Licensee MDPI, Basel, Switzerland. This article is an open access article distributed under the terms and conditions of the Creative Commons Attribution (CC BY) license (http://creativecommons.org/licenses/by/4.0/).

MDPI  
St. Alban-Anlage 66  
4052 Basel  
Switzerland  
Tel. +41 61 683 77 34  
Fax +41 61 302 89 18  
www.mdpi.com

*Crystals* Editorial Office  
E-mail: crystals@mdpi.com  
www.mdpi.com/journal/crystals

www.ingramcontent.com/pod-product-compliance
Lightning Source LLC
LaVergne TN
LVHW071955080526
838202LV00064B/6758